Springer Polar Sciences

Series Editor

James D. Ford, Priestley International Centre for Climate, University of Leeds, Leeds, West Yorkshire, UK

Springer Polar Sciences is an interdisciplinary book series that is dedicated to research in the Arctic, sub-Arctic regions, and the Antarctic. In recent years, the polar regions have received increased scientific and public interest. Both the Arctic and Antarctic have been recognized as key regions in the regulation of the global climate, and polar ecosystems have been identified to be particularly susceptible to the ongoing environmental changes. Consequently, the international efforts in polar research have been enhanced considerably, and a wealth of new findings is being produced at a growing rate by the international community of polar researchers.

Springer Polar Sciences aims to present a broad platform that will include state-of-the-art research, bringing together both science and humanities to facilitate an exchange of knowledge between the various polar science communities. The Series offers an outlet to publish contributions, monographs, edited works, conference proceedings, etc. Topics and perspectives will be broad and will include, but not be limited to: climate change impacts, environmental change, polar ecology, governance, health, economics, indigenous populations, tourism and resource extraction activities. Books published in the series will appeal to scientists, students, polar researchers and policy makers.

More information about this series at https://link.springer.com/bookseries/15180

Jeffrey McGee · David Edmiston · Marcus Haward

The Future of Antarctica

Scenarios from Classical Geopolitics

Springer

Jeffrey McGee
Faculty of Law
Institute for Marine and Antarctic Studies
University of Tasmania
Hobart, TAS, Australia

David Edmiston
Faculty of Law
University of Tasmania
Hobart, TAS, Australia

Marcus Haward
Institute for Marine and Antarctic Studies
University of Tasmania
Hobart, TAS, Australia

ISSN 2510-0475 ISSN 2510-0483 (electronic)
Springer Polar Sciences
ISBN 978-981-16-7094-7 ISBN 978-981-16-7095-4 (eBook)
https://doi.org/10.1007/978-981-16-7095-4

This Springer imprint is published by the registered company Springer Nature Singapore Pte Ltd.
The registered company address is: 152 Beach Road, #21-01/04 Gateway East, Singapore 189721, Singapore

Foreword

Antarctica is a magnificent place, a vast area of the planet that has been free from large-scale nation-state conflict from the advent of the age of exploration, when pioneers like Amundsen, Bellingshausen, Byrd, Mawson, Scott, Shackleton and Shirase arrived, to the present day. When the Cold War threatened to extend global tensions to this region, twelve nations negotiated the Antarctic Treaty of 1959, effectively setting aside differences over territorial claims and preventing militarization over a significant part of the globe. Over time, with participation of many more states, additional instruments covering environmental protection and marine conservation were added to create the Antarctic Treaty System. The intention by these states was to establish a "natural reserve, devoted to peace and science", and as a general proposition, the Treaty System has been successful and that vision has been achieved.

Yet the world is experiencing considerable geopolitical turmoil, and Antarctica is not immune. While governance in Antarctica has gone smoothly by comparison with other regions, the underlying circumstances of relations among states in Antarctica have evolved over the past six decades. There are any number of economic and security tensions in Antarctica that require careful attention so that a peaceful future for the region can be secured. There is a growing desire to know: Is the political future of the region stable? How will states balance economic (and in particular fisheries) interests? Can the status quo with regard to territorial claims continue? How will climate change affect the interests of and actions of states in the Antarctic? Will the current ban on mining continue perpetually? Can the region remain unmilitarized?

Given these critical questions, this new book, from a series of leading experts, comes at just the right time. The authors, prominent scholars at the University of Tasmania, located in Hobart, Australia's gateway to Antarctica and home to the Australian Antarctic Division, have developed original scholarship with the goal of providing a new perspective on the prospects for political stability in the Antarctic region. Using a multidisciplinary approach, they have systematically considered how the future of the Antarctic Treaty System may develop, using a series of scenarios, and focusing in particular on the question of militarization. They present background and history on the Antarctic Treaty System, describe aspects of scenario planning, focus on detailed aspects of militarization and ultimately construct five scenarios on the issue of Antarctic militarization. Along the way, they focus on important topical issues such as the impact of the rise of China on Antarctic geopolitics and other aspects of Great Power influence, the role of territorial claimants in Antarctic geopolitics and current and future interests in resources that could upset the current order.

The result is a comprehensive consideration of possible future paths. The analysis will be valuable to diplomats, political scientists and government officials, as well as all those interested in whether Antarctica will be able to meet the international community's expectation that it will retain its dedication to peace and science.

Evan T. Bloom
Former lead for U.S. Antarctic
Diplomacy (2006–2020)
Senior Fellow, Woodrow Wilson
International Center for Scholars
Washington, US

Preface

This book is written from the Antarctic gateway city of Hobart, Tasmania—a vibrant harbour for scientific and other interest in Antarctica in the Southern Ocean. Hobart played an important part in the "Heroic Era" expeditions of key figures such as Roald Amundsen and Sir Douglas Mawson. It is also home to Australia's National Antarctic Program, administered by the Australian Antarctic Division, and much of the Antarctic scientific expertise of Australia's national scientific body, the CSIRO. Over summer, it is usual to see icebreakers from the Australian, French and Chinese national Antarctic programs leaving port for Antarctic research stations in the south. Each October, the Member States in the Commission for the Conservation of Antarctic Marine Living Resources (CCAMLR) travel to Hobart for a two-week meeting on marine resource management in the Southern Ocean. In ordinary times, there is a regular flow of people into Hobart for presentations on Antarctic science, policy, humanities and the arts. The temperate, windy and unpredictable weather here reminds us that Antarctica and the Southern Ocean are our neighbours to the south. They have a palpable impact on the mood and feel of the region. The authors of this book share an appreciation of the history and connection between Tasmania, Antarctica and the Southern Ocean.

All three authors are located within the Institute for Marine and Antarctic Studies (IMAS) or the Faculty of Law at the University of Tasmania. This book is an inter-disciplinary collaboration in that it draws together authors from the disciplines of international law, international relations and public policy to explore the futures of Antarctica through a common lens of classical geopolitics. Our shared interest in the future of the Antarctic Treaty System (ATS) is longstanding: Haward first published on the topic in 1990—at a time when the ATS faced significant pressures as it pivoted from a focus on regulating mining towards environmental protection in Antarctica. The more immediate inspiration for this book comes from a coursework masters project on scenario analysis and Antarctic futures that Edmiston conducted in 2019. This explored how the general literatures on scenario analysis might be applied to assess geopolitical futures for Antarctica.

That work was expanded and refined in this book and is illustrated through a deep engagement with the topical issue of Antarctic militarization. Other case studies might have been chosen, such as mineral resources or fisheries. In selecting militarisation as our focus, we do not wish to add weight to the more speculative claims on this topic that are already in the public sphere. Rather, we have sought to approach the issue in a rigorous, evidence-based and analytical manner, relying on information that is available in the public sphere and validating our conclusions with a small number of informed experts. We hope these scenarios do not incite undue alarm. Our aim is simply to highlight some of the pressures that might lead, plausibly, to a weakening of the historically successful non-militarization provisions of the Antarctic Treaty. We hope this volume encourages government agencies and researchers from various disciplines to consider the use of rigorous scenario approaches and techniques to inform their analyses of the future of the ATS. In doing so, we also hope to promote the resilience of the ATS and further its impressive record of managing the region over the last 60 years.

Hobart, TAS, Australia
August 2021

Jeffrey McGee
David Edmiston
Marcus Haward

Acknowledgments

The authors wish to thank the Australian Research Council for its support of the book through Discovery Grant (DP19) “Geopolitical Change and the Antarctic Treaty System” and for access to documents from the Linkage Infrastructure Equipment Facilities (LIEF18) Grant “Creating a Unique Open Access Database on Antarctic Law and Governance”. Aspects of this research were supported by an Australian Government Research Training Program (RTP) Scholarship.

The authors acknowledge the generous contributions of colleagues in Hobart, Canberra, Washington, D.C., and Rotterdam, who have been kind enough to act as sounding boards for our ideas, give feedback on drafts or provide ideas for further enhancements. In particular, we would like to thank (in alphabetical order) Mr. Bruno Arpi, Dr. Anthony Bergin, Mr. Evan Bloom, Mr. Malcom Davis, Prof. John Dickey, Dr. Leonhardt van Efferink, Dr. Andrew Jackson, Profs. A. J. Press, Shirley Scott and Ms. Claire Young. Any errors or omissions of course remain the sole responsibility of the authors.

We would also like to thank Dr. Jacqueline Fox for her excellent copyediting assistance, which made the final stages of submission of this book a relatively painless, and even enjoyable, experience. Many thanks also go to Dr. Stacey McCormack for laying out our diagrams in her usual expert fashion.

Jeffrey McGee would like to thank his family, wife Maree and children Connor, Lucas and Declan, for their understanding and cheerful support during the long hours required to finalize this volume. David Edmiston is very grateful to his wife Gillian and children Hilary and Oliver. He apologizes to Otto, who missed out on so many walks. Marcus Haward wishes to thank Anne and Charles for their ongoing support.

Praise for *The Future of Antarctica*

"This timely and readable volume is social science at its best insofar as its authors make an original and valuable contribution to our understanding both of the subject at hand and to methodology. Applying scenario analysis to the issue of the militarization of the continent, the authors engage critically with the logic of the approach to yield methodological insights that will be helpful to organizations and educators who use scenario-based methodologies. The policy conclusions will be of interest to all those committed to a peaceful future for the Antarctic continent."

—Shirley Scott, *Professor of International Law and International Relations Head, School of Humanities and Social Sciences, UNSW Canberra at the Australian Defence Force Academy*

"This volume provides an outstanding contribution to the understanding of present and future Antarctic politics. By using scenario analysis, McGee, Edmiston and Haward give us a careful and thought provoking vision of Antarctica's next decade, where climate change, balance of power and technological development might make the region a critical component of world politics. This book's strength lies in aligning classic geopolitics with robust methodological frameworks, smoothly describing Antarctica's realities and key drivers, and providing the reader with plausible consequential scenarios—some of which can be quite surprising. This is a must-read for both academics and policy-makers who seek to understand how Antarctic geopolitics might develop in the near future."

—Dr. Daniela Portella Sampaio, *Faculty of Sociology, Bielefeld University, Germany*

Contents

1 **Introduction** 1
- 1.1 The Antarctic Treaty System: Responsive, Resilient—And Largely Successful 2
- 1.2 The Antarctic Treaty System: Fears for the Future? 4
- 1.3 Antarctic Futures and Scenarios 5
- 1.4 Antarctic Futures: A Multi-disciplinary Approach 7
- 1.5 Classical and Critical Geopolitics 8
- 1.6 Towards a Critical Realist Geopolitics? 10
- 1.7 What This Book is About 11
- References 13

2 **History of Antarctic Territorial Claims and Spatial Contestation** 17
- 2.1 Introduction 17
- 2.2 The 'Antarctic Problem' 18
- 2.3 Geopolitics and Resolving the 'Antarctic Problem' 20
- 2.4 The Antarctic Treaty: Setting Aside Geopolitical Differences? 23
- 2.5 Geopolitics and the Future of Antarctica 27
- References 28

3 **Scenario Analysis and the Limits of Prediction** 31
- 3.1 Overview 31
- 3.2 What Are Scenarios? 32
- 3.3 Ten Key Properties of Scenarios 35
- 3.4 Reality Check: Scenarios and Our Need to Predict 39
- 3.5 Plausibility and the 'aha' Moment 40
- 3.6 A Revised Model: Introducing the Zone of Plausibility 42
- 3.7 Dissecting Turbulence, Uncertainty, Novelty and Ambiguity (TUNA) 44
- 3.8 Summary 45
- References 47

4 Objectives, Approaches and Techniques 51
4.1 Introduction 51
4.2 Objectives, Approaches, and Techniques 53
4.3 Framing Before Reframing 54
4.4 Level 1—Define Your Strategic Objectives 55
4.5 Level 2—Identify Your General Approach/es 55
4.6 A Scenario Toolbox 56
4.7 Deductive Approaches: Intuitive, Archetypal, and Normative 57
4.8 Deductive (Intuitive) Approaches 58
4.9 Deductive (Archetype) Approaches 59
4.10 Deductive (Normative) Approaches 60
4.11 Inductive Approaches 60
4.12 Quantitative approaches 61
4.13 Hybrid approaches 63
4.14 Integrated approaches 64
4.15 Scoping approaches 65
4.16 Engagement approaches 65
4.17 Level 3—Select Your Specific Technique/s 69
4.18 The 'Intuitive Logics 2 × 2' Technique 69
4.18.1 Step 1: Identify Your Scenario Standpoint 69
4.18.2 Step 2: Set Your Time Horizon 70
4.18.3 Step 3: Identify the Key Driving Forces 70
4.18.4 Step 4: Distinguish Critical Uncertainties from Predetermined Elements 71
4.18.5 Step 5: Rank Critical Uncertainties by Importance and Uncertainty 72
4.18.6 Step 6: Select Your General Approach/es 72
4.18.7 Step 7: Construct the Scenario Matrix 72
4.18.8 Step 8: Develop the Scenario Narratives 75
4.18.9 Step 9: 'Wind Tunnel Test' Current Strategy and Prepare Policy Advice 75
4.19 Insights from the Process and Adding Depth: An Alternative to IL 2 × 2—Towards 'IL 5' 76
4.20 Summary 78
References 79

5 Antarctic Geopolitics: Background 83
5.1 Overview 83
5.2 A Brief Recap: Geopolitics—Classical, Critical, and Other Approaches 84
5.3 Antarctica and Classical Geopolitical Analysis 85
5.4 Nine Themes for the Driving Forces of Antarctic Geopolitics 86
5.5 Challenges for the Scenario Planner! 87

5.6 Four Dynamics of Driving Forces ... 89
5.7 Pulling These Together: The 'Dynamic Signature' of a Driving Force ... 92
5.8 Nine Themes for Driving Forces ... 93
5.9 Summary ... 99
References ... 100

6 Militarisation of Antarctica ... 105
6.1 Introduction ... 105
6.2 History of Militarisation in Antarctica ... 106
6.3 Managing Militarisation Under the Antarctic Treaty 1959 ... 111
6.4 Traditional Military/Strategic Threats in Antarctica ... 112
6.5 'Dual-Use' Equipment in Antarctica ... 119
6.6 Summary ... 122
References ... 123

7 Antarctic Militarisation: Scenario Analysis ... 125
7.1 Introduction ... 125
7.2 Step 1: Identify Your Scenario Standpoint ... 126
7.3 Step 2: Set Your Time Horizon ... 127
7.4 Step 3: Identify Driving Forces of Antarctic Militarisation ... 127
7.5 Step 4: Classify Driving Forces as Predetermined Elements or Critical Uncertainties ... 129
7.6 Step 5: Rank Driving Forces by Importance and Uncertainty ... 138
7.7 Step 6: Select Your General Approach/es ... 144
7.8 Step 7: Construct the Scenario Matrix ... 145
7.9 Summary ... 149
References ... 150

8 Antarctic Militarisation: Five Scenarios ... 153
8.1 Introduction ... 153
8.1.1 Scenario 1: Splendid Isolation ... 154
8.1.2 Scenario 2: Quarantine ... 155
8.1.3 Scenario 3: Stealth ... 157
8.1.4 Scenario 4: Cold War II ... 158
8.1.5 IL 5: A Fifth Scenario for Antarctic Militarisation ... 161
8.1.6 Scenario 5: Pressures Managed ... 162
8.2 Step 9: Wind Tunnel Test Current Strategy and Prepare Policy Advice ... 164
8.3 Summary ... 166
Reference ... 167

9 Scenario Analysis and the Classical View of Antarctic Geopolitics 169
9.1 Introduction 169
9.2 Geopolitics: The Intersection of Power, Authority, and Space 170
9.2.1 Power 170
9.2.2 Military Power 171
9.2.3 Scientific and Logistical Power 172
9.2.4 Authority 174
9.2.5 Space 175
9.3 Antarctica Through a Classical Geopolitical Lens 176
9.3.1 Power and International Law 178
9.3.2 Antarctica and 'Space in Rivalry Between States' 180
9.4 Summary 182
References 183

10 Conclusion 185
10.1 Our Objectives 186
10.2 Our Findings: Geopolitical Scenario Analysis 187
10.3 Our Findings: Geopolitical Futures for Antarctica 189
References 192

Index 193

About the Authors

Jeffrey McGee is an associate professor at the Institute for Marine and Antarctic Studies (IMAS) and Faculty of Law at the University of Tasmania. His work is published in leading international journals in the fields of Antarctic policy, international environmental law and climate change policy. He co-edited the book *Anthropocene Antarctica*, a special issue of the *Australian Journal of Maritime and Ocean Affairs* on Twenty-First-Century Challenges to the Antarctic Treaty System, and the *Edward Elgar Research Handbook on Climate Change, Oceans and Coasts.* He is an affiliated researcher with Humanities and Social Science expert group of the Scientific Committee on Antarctic Research. He is also a member of the Australian Government's consultative forum for the Convention on the Conservation of Antarctic Marine Living Resources and the Tasmanian Polar Network. In 2021, Jeff was a member of the Australian delegation to the 43rd Antarctic Treaty Consultative Meeting. He also has experience as a lecturer on tourist flights to the Ross Sea area and East Antarctica.

David Edmiston is a Ph.D. candidate in the Faculty of Law at the University of Tasmania. A career public servant, David holds masters degrees in public administration and environmental governance, with a focus on Antarctica. In 2019, David undertook an intensive course in geopolitical scenario planning at the Maastricht University Summer School, under the guidance of Dr. Leonhardt van Efferink. David's doctoral project is titled: *Portents from the North: Is the Arctic an indicator of future challenges for Antarctic geopolitics, law and governance?*

Marcus Haward is the professor of Antarctic and Ocean Governance at the Institute for Marine and Antarctic Studies, University of Tasmania. He has been a member of Australian delegations to the Antarctic Treaty Consultative Meeting and to the Commission for the Conservation of Antarctic Marine Living Resources. He has published extensively on Antarctic and ocean governance. He is currently co-leading (with Jeffrey McGee) a multi-year project on Antarctic Geopolitics funded by the

Australian Research Council. He co-edited *Australia and the Antarctic Treaty System: Fifty Years of Influence* UNSW Press, 2011 and most recently authored *Governing Oceans in a Time of Change: Fishing for a Future*, Edward Elgar, 2020.

Abbreviations

AAT	Australian Antarctic Territory
ATCM	Antarctic Treaty Consultative Meeting
ATCP	Antarctic Treaty Consultative Party
ATS	Antarctic Treaty System
CAMLR	(Convention for the) Conservation of Antarctic Marine Living Resources 1980
CCAMLR	Commission for the Conservation of Antarctic Marine Living Resources
CCAS	Convention for the Conservation of Antarctic Seals 1972
CCCR	Command, Control, Communications and Reconnaissance
CEP	Committee for Environmental Protection
COMNAP	Council of Managers of National Antarctic Programs
CRAMRA	Convention for the Regulation of Antarctic Mineral Resource Activities 1988 (not in force)
CTBT	Comprehensive Nuclear Test Ban Treaty 1996
CU	Critical Uncertainty
IAATO	International Association of Antarctica Tour Operators
ICRW	International Convention for the Regulation of Whaling 1946
IL	Intuitive Logics
IMO	International Maritime Organization
IUU	Illegal, Unreported and Unregulated (fishing)
MARPOL	International Convention for the Prevention of Pollution from Ships 1973
MPA	Marine Protected Area
NATO	North Atlantic Treaty Organization
NCP	Non-Consultative Party (Antarctic Treaty)
OSPA	Oxford Scenario Planning Approach
PESTEL	Political, Economic, Social, Technological, Environmental and Legal
SCAR	Scientific Committee on Antarctic Research

TUNA	Turbulence, Uncertainty, Novelty and Ambiguity
UNCLOS	United Nations Convention on the Law of the Sea 1982
UNGA	United Nations General Assembly

List of Figures

Fig. 2.1 Antarctic territorial claims. *Source* AAD Data Centre. https://data.aad.gov.au/database/mapcat/antarctica/territ orial_claims_b&w_13567.pdf 19
Fig. 3.1 Limits of prediction due to complexity, uncertainty and the passage of time 34
Fig. 3.2 Zones of the predictable, the plausible and the possible 44
Fig. 4.1 Broad scenario approaches 57
Fig. 4.2 Scenario matrix 73
Fig. 4.3 Two critical uncertainties 74
Fig. 4.4 Scenario matrix—worked example 74
Fig. 4.5 Two critical uncertainties at three levels (high, medium and low) 77
Fig. 5.1 Nine themes for driving forces: PESTEL + three dsiciplines of geopolitics 88
Fig. 5.2 Dynamics of driving forces in Antarctic geopolitics 92
Fig. 6.1 Sites of Military Activity in the Antarctic Region prior to 1959 110
Fig. 7.1 Dynamic signature of CU 1 at and low and high instances 146
Fig. 7.2 Dynamic signature of CU 2 at low and high instances 147
Fig. 7.3 Four scenarios of Antarctic militarisation 148
Fig. 8.1 An 'IL 5' model for Antarctic militarisation 161

List of Tables

Table 2.1 Antarctic territorial claims and dates of claim 18
Table 2.2 Number of parties to the Antarctic Treaty by decade of accession 25
Table 4.1 Scoping options for scenarios 66
Table 7.1 Driving forces for militarisation in Antarctica 130
Table 7.2 Scoring the driving forces for militarisation in Antarctica 139
Table 7.3 Top two critical uncertainties for Antarctic militarisation 143
Table 7.4 Other significant critical uncertainties for Antarctic militarisation 144
Table 7.5 Top two critical uncertainties for the scenario matrix 145
Table 7.6 An IL 2 × 2 scenario logic for Antarctic militarisation 147

Chapter 1
Introduction

Abstract There is growing interest in the future of Antarctic governance. For the last 60 years, the Antarctic Treaty System (ATS) has been comparatively successful in meeting its objectives, and sufficiently adaptive in the face of change. However, within the literature, there is increasing concern that, in the medium to long term, ATS responses to emerging geopolitical challenges may be inadequate. How do we assess this concern? How can organisations evaluate the future robustness of their current strategy? Antarctic futures have been explored within national Antarctic programs and they have been the subject of serious academic attention. Unfortunately, much of this commentary lacks a common language and analytical approach. To address this gap, we explore how the multi-disciplinary focus of classical geopolitics might be applied systematically to Antarctic scenarios. Here, we define geopolitics as the intersection between power, authority, and space, in rivalry between states. Over the course of this book, we illustrate a possible approach with the topical example of Antarctic militarisation. Along the way, we make suggestions to augment the current theory and practice of geopolitical scenario planning. We hope this work may be useful for those interested in the rigorous assessment of geopolitical futures—in Antarctica and beyond.

Keywords Antarctic futures · Antarctic scenarios · Geopolitical scenario planning · Classical geopolitics · Critical geopolitics · Critical realist geopolitics · Ggeopolitics definition

There is no shortage of popular and academic interest in the future of Antarctica. Its remoteness and the challenging physical conditions for human existence are a source of continuing fascination with the ice-covered continent. This interest extends to the current and future political status of Antarctica, and the governance arrangements in place under the Antarctic Treaty System (ATS). Commentators on Antarctic affairs routinely speculate on the impact of current activities for the geopolitical future of the continent and for the ATS. This commentary is varied and at times takes some surprising and eye-opening turns. For instance, in recent years some writers have called for Australia to pay close attention to how it might militarily defend its Antarctic territorial claim; some have expressed alarm at the expansion of Chinese activities in Antarctica, and even suggested that New Zealand look to Antarctica as an

J. McGee et al., *The Future of Antarctica*, Springer Polar Sciences,
https://doi.org/10.1007/978-981-16-7095-4_1

area of *Lebensraum* (Liggett et al. 2017)—a concept deployed from the early 1900s through to World War II to support arguments for 'natural' territorial expansion of the German state. While this commentary provokes significant public interest, it is important for the future of Antarctic governance that geopolitical assessments be approached in a systematic and rigorous manner.

Affairs between states are more difficult to govern than affairs between citizens, government, corporations, and civil society at the domestic level. The domestic state creates legal rules and applies coercive measures to ensure reasonable compliance. At the international level, however, a state has significantly less capacity to bring coercive measures to bear upon another state to encourage compliance with the rules of international law (Alonso-Trabanco 2019). The international system has no world government or sovereign with a legislative and enforcement function akin to that of the domestic state. Instead, the international system relies on a range of incentives and disincentives that states can use to encourage other states to comply. It is this cooperation between sovereign states in the international system that lies at the heart of international law (Keohane and Nye 1973). Over time, these rules can develop normative force (Pettit 2019) which creates reputational risks for non-compliant states (Guzman 2008).

1.1 The Antarctic Treaty System: Responsive, Resilient—And Largely Successful

By any objective measure, the ATS is one of the good news stories of international governance and law (Young 2010). Since the late 1950s, the area of below 60° south has been effectively non-militarised, made off-limits for nuclear testing or storage of nuclear weapons. It has been used in a manner which is unusual for nation states; that is, as an open area for scientific investigation and other peaceful use. As we discuss in Chap. 2, the Antarctic Treaty of 1959 was established primarily as a peace agreement to diffuse possible international conflict over the Antarctic continent amid heightened international tension from the Cold War. The actions of the 12 original signatories to the Antarctic Treaty were a remarkable achievement. They successfully negotiated an agreement in this Cold War context, which effectively created a zone of peaceful use for seven per cent of the planet. During the 60 years the Treaty has been in force, there has been no compelling evidence of any activities of a military nature being carried out in the area below 60° south. Concerns have been raised about the possibility of a 'dual use' of some personnel and equipment for both peaceful and military purposes—as we discuss in Chap. 6. Even with the military conflict between the United Kingdom and Argentina in the Falkland Islands/Islas Malvinas war of the early 1980s, there were no obvious military activities carried out in the Antarctic Treaty Area (ATA) itself during this period of heightened tension.

As we see in Chap. 2, not only has the Antarctic Treaty been highly successful in terms of ensuring that the area below 60° south is used only for peaceful purposes; the

Treaty has also seeded development of an interconnected group of further treaties and binding legal measures, in response to wider governance concerns. Within a couple of years of the Antarctic Treaty coming into force in 1961, the 12 original signatories and consultative parties formed binding legal measures to manage concerns about human impacts upon flora and fauna on the Antarctic continent. The 1963 Agreed Measures for the Conservation of Antarctic Flora and Fauna (Agreed Measures) provided a regime of environmental protection for the Antarctic continent, particularly directed towards managing the activities of research bases on the continent and its ice shelves.[1]

Similarly, during the early 1970s, the Antarctic Treaty Consultative Parties initiated negotiation of a new convention to avoid depletion of seal populations in the Antarctic Treaty area, which had been decimated during the nineteenth century. The Convention for the Conservation of Antarctic Seals (CCAS), concluded in 1972, has remained in place as a governance regime ready to respond to any reinvigoration of sealing activity in Southern Ocean areas below 60° south. Further arrangements for managing the resources of the Southern Ocean were agreed in 1980 through the Antarctic Treaty Consultative Meeting (and initiating negotiations). This led to the Convention on the Conservation of Antarctic Marine Living Resources (CAMLR Convention), which came into effect in 1982. Through progressive implementation of concepts such as precaution and ecosystem-based management, the Commission for the Conservation of Antarctic Marine Living Resources (CCAMLR) has successfully managed the finfish and krill harvesting activities of states in the Southern Ocean and established a conservation regime for the region.

During the 1980s, the Antarctic Treaty Consultative Meeting initiated negotiations on the Convention for the Regulation of Antarctic Mineral Resource Activities (CRAMRA). This was due to concern by some states that the potential for mineral prospecting, exploration and extraction exposed a significant governance gap in the ATA. Although mining in the Antarctic continent and the offshore areas of the Southern Ocean was not anticipated to occur in the immediate future, some states (such as New Zealand) took the view that it was important to have in place a rigorous set of legal rules to govern any future mineral resource activities—including environmental assessments to inform decision-making. The text of the CRAMRA Convention was agreed in 1988. Soon after, however, Australia and France indicated that they would not sign the convention and instead embarked upon a diplomatic initiative to form a treaty for Antarctic environmental protection—as discussed in Chap. 2. It is worth noting here that the resultant Environmental Protocol to the Antarctic Treaty (the Madrid Protocol) was agreed by the Antarctic Treaty Consultative Parties in 1991. Since that time, the parties have constructed a detailed set of rules to protect the environment of the Antarctic continent and ocean areas within the ATA. These rules now form a central part of the objectives and activities of the Antarctic Treaty System.

[1] The Agreed Measures were superseded with the entry into force of the Protocol on Environmental Protection to the Antarctic Treaty in 1998 (the Madrid Protocol) and were finally extinguished by the Antarctic Treaty Consultative Parties in 2011.

The adaptability and responsiveness of the Antarctic Treaty Consultative Parties is, in many ways, a remarkable achievement. In addition, membership of the Antarctic Treaty has expanded from its original 12 members to 54, and now encompasses over 80 per cent of the global population (Natural Environment Research Council (NERC)/British Antarctic Survey (BAS) 2015). The CCAMLR membership has also grown to 26 states (plus the European Union) and includes major distant water fishing countries and regional groupings. Issues with illegal, unreported, and unregulated (IUU) fishing within the CCAMLR area during the early 2000s have largely been dealt with through innovative approaches, such as catch documentation and satellite monitoring of vessels. Overall, levels of compliance with, and confidence in, the Antarctic Treaty and its associated instruments remains fairly high (Young 2010).

1.2 The Antarctic Treaty System: Fears for the Future?

A significant strand of academic analysis nonetheless portrays the ATS as beset by major challenges and issues. Fears are expressed that the ATS may be unable to respond adequately to these challenges, or that major changes will be forced upon it (Hemmings 2014; Dodds and Nutall 2016). In particular, there is ongoing speculation about the future intent of China, in the context of its expanding number of Antarctic research bases (to five) and its interest in commercial krill fisheries. China has also been identified as a potential future architect of a significant challenge to the ATS when the Madrid Protocol comes up for possible review from 2048 onwards. Significant concern is raised by scholars such as Brady (Brady 2017) and Abdel-Motaal (Abdel-Motaal 2016) that China may be positioning itself to overturn the Madrid Protocol's current ban on mineral resource activities. Similarly, Hemmings (Hemmings et al. 2017) points to future Antarctic challenges driven by climate change, increased resource demands, commercialisation, and political pressures over the dormant (but unresolved) issue of claims to territorial sovereignty. While hopefully a more transient issue, recent commentary has also drawn attention to the difficulties posed by the COVID-19 pandemic for Antarctic research activities and governance of the continent (Frame and Hemmings 2020).

We are sympathetic to much of this work. It seeks to identify likely drivers of change within the Antarctic region and to the ATS, in a world where dynamic physical systems are undergoing accelerated perturbations due to global climate change, and where the international system is shifting back from a unipolar to a multipolar power structure. We agree that it is important not to be blinded by the past success of institutional patterns. We agree it is crucial for the analyst to be open to the potential impacts of new drivers and emerging patterns of non-linear biophysical and institutional change. However, we also argue that a methodical, systematic approach to analysis of the future of Antarctica and the ATS is important. Why? Because it can provide a common language, methodology and set of assumptions for a constructive dialogue on Antarctic futures between policy makers, academics and civil society.

1.3 Antarctic Futures and Scenarios

Scenario planning, in various guises, underpins significant research and logistics activity conducted by national programs in the Antarctic and the Southern Ocean. For example, Australia has used scenario frameworks in key planning exercises, including the Australian Antarctic Science Advisory Committee's Foresight Report (1997) (Antarctic Science Advisory Committee (ASAC) 1997). This was pivotal in shaping strategic assessments and formulating the Australian Government's goals for the Australian Antarctic Program in the mid-2000s [Australian Antarctic Division, unpublished]. Within the broader ATS, we are aware that scenario analysis has been used by the Scientific Committee on Antarctic Research (SCAR). Conducted in 2014, SCAR's *2020Horizon Scan* (Kennicutt et al. 2014a, b) was a major assessment of future science issues and challenges that included a 'roadmap' to address them.

Antarctic futures have also been the subject of serious academic attention. Notable forward-looking contributions include Powell and Dodds' edited book on polar geopolitics (2014) (Powell and Dodds 2014; Pincus and Ali 2015; Abdel-Motaal 2016; Dodds and Nutall 2016; Brady 2017; Wehrmann 2019). The potential for contestation over Antarctic resources is a recurrent concern. More formal scenario analysis has also been employed in Antarctic research (Haward et al. 2003). Prominent examples include Tin and colleagues' edited volume on Antarctic futures (2011), with its focus on environmental issues (Tin et al. 2014). From 2017, Liggett and colleagues' paper on Antarctic futures to 2040 is substantial and well-reasoned, albeit focusing on a subset of possible topics (Liggett et al. 2017). More recently, Frame (2019, 2020) elaborates a typology and framework for Antarctic futures, and notes that 'there is currently no coordinated approach to establish a workable set of scenarios in which future possibilities could be actively explored' (Frame 2020, p. 22). We concur and suggest that modelling of Antarctic's *geopolitical* futures is not nearly as comprehensive, intensive, or well-resourced as it is for its *environmental* futures—see, for example, the Intergovernmental Panel on Climate Change's *Special Report on the Ocean and Cryosphere in a Changing Climate* (2019) (Intergovernmental Panel on Climate Change (IPCC) 2019). Given what may be at stake, we suggest this is a void that needs to be filled.

A recent, publicly available example of government scenario planning for Antarctica is the Royal Australian Air Force's report, 'Australia's National Antarctic Airpower Futures' (Layton et al. 2019). This draws on a broader 2016 report in which scenario analysis was used by the Australian Defence Force to analyse its future operating environment at Environment and 2035 (2035). Also in 2019, the Australian Government's Civil-Military Centre published *Antarctica 2050: Strategic Challenges and Responses* (Press et al. 2019). This brought together discussion papers by prominent Australian and New Zealand academics on future challenges to the ATS. A recurrent concern is ensuring the resilience and adaptability of the ATS to meet emerging challenges. These include the role of science as a strategic activity, interest in Antarctica's resources, and the region's growing geostrategic importance.

More recently, the sixtieth anniversary of the coming into effect of the Antarctic Treaty System in 2021 saw the publication of several analyses of the future of the ATS. Frame and colleagues imagine two scenarios out to 2050, set in a post-COVID-19 world riven by a fragmenting global order and the impacts of climate change (Frame et al. 2021). The key point of divergence between their 'dystopian' and 'utopian' scenarios is the degree to which the rules and norms of the international system are upheld. The primary role of these scenarios is to advance a normative (i.e., critical theory inspired) vision for the future of Antarctic governance. The authors are careful to position this exercise as 'informal', 'performative' and possibly 'fanciful', rather than systematic (Frame et al. 2021, pp. 1–3). Of course, no one can challenge scenarios of this sort, because it would be like arguing with a poem.

Further, a recent strand of policy and think tank literature brings a security framing to Antarctic futures, without engaging in scenario analysis. On the anniversary of the Antarctic Treaty, Buchanan (Buchanan 2021) suggests a range of potential challenges facing the Antarctic Treaty System over the sixty years to 2081. In a similar vein, Strating and Buchannan (2021) argue that the growing geopolitical focus on the future of the Indo-Pacific region needs to include analysis of competitive state behaviour that is allegedly already occurring within the ATS. Specifically, they identify certain 'grey zone activities' allegedly occurring in the ATA, which they define as states 'working the system through such measures as dual-use technologies (i.e., those which have both scientific and military applications) and frustrating ATS consensus mechanisms (blocking environmental management plans)' (Strating and Buchanan 2021). Finally, we note that the United States Modern War Institute at West Point has established 'Project 6633', which broadly seeks to provide an international collaboration to assess 'the future security trajectories of the Arctic and Antarctica' (Buchanan and Burke 2021).

The above literatures have made an important contribution to spurring thinking about Antarctic futures and expanding the disciplinary perspectives through which they are developed. However, it appears that the ability (or inclination) to undertake formal scenario analysis is seldom found among experts in geography, geopolitics, polar governance, and law. In this book, we contend that a well-designed and executed scenario planning process can both fire the imagination *and* be firmly tethered in a contemporary evidence base, with rich data analysis and robust chains of causal reasoning. Critically, much of the richness (and most valuable insight) is generated from a highly iterative, intense, and imaginative set of processes. Successful use of scenario planning depends on a well-planned, systematic approach. Collaborative projects that combine various disciplinary specialisations with skilled scenario practitioners could yield powerful analytical results. In providing a window into plausible alternative futures, such scenario work may validate or challenge current policy settings.

1.4 Antarctic Futures: A Multi-disciplinary Approach

This volume is intended to have multi-disciplinary appeal. Traditionally, academic work on the future of Antarctic governance has been taken up by three bodies of literature with distinct disciplinary appeals. First, *international legal analysis* of the ATS has predominantly adopted a legal doctrinal methodology (Joyner and Chopta 1988; Triggs 1986,1987). This focuses on explication of legal rules and norms within the various instruments of the ATS, as well as norms of non-state governance developed by bodies such as the International Association of Antarctica Tour Operators (IAATO). International legal literature therefore focuses upon the content and interpretation of various rules within the instruments of the ATS. It also seeks to identify ambiguities and gaps that might be resolved, with a view to enhancing the normative ordering of state behaviour. The underlying assumption of much of this literature is that 'new and/or more rules' can be a solution to many governance problems in Antarctica. Legally inspired futures analysis therefore has tended to focus on the extrapolation of existing rules, and on futures based on alternative normative ordering. However, we note that over the last few decades, purely legal analysis of Antarctic affairs has diminished, with the last major books published in the early 2000s. Legal analysis of Antarctic futures is now more likely to be found within broader interdisciplinary (McGee and Liu 2019) or themed projects (Dodds et al. 2017).

Second, over the past forty years, *institutionalist* literature (emanating from the disciplines of political science and international relations) has developed a rich and extensive body of work on 'regime theory'. This provides an explanatory framework on the formation, effectiveness, legitimacy, and problem-solving attributes of international governance. Regime theory has been adapted by different authors to provide interest-based analyses of various aspects of Antarctic governance. These include Young Oran 1993; Stokke and Vidas 1996; Joyner 1988; Young 2010; McGee and Haward 2019). This work focuses on the role of state interests and the capacity of international cooperation and institution building to overcome collective action problems. Application of scenario analysis to regime theory would therefore focus on the capacity of institutions (and their members) to overcome their current governance gaps and their present failures of cooperation in international spaces, in order to tackle plausible future challenges.

However, the literatures of international law and regime theory engage only indirectly with key issues such as the role of states (and other actors) in the creation of international problems, and the capacity for law and institutions to respond with effective solutions. For this, we turn to our third discipline in the literature relating to the future of Antarctic governance: *geopolitics*. This is concerned with 'the relation of international political power to the geographical setting' (Cohen 1973, p. 25) and seeks to make explicit the broad connection between power, spatial ordering, and institutions. While 'geopolitics' is used ubiquitously in popular and academic debate, as with many concepts, the discipline has not yet settled on a widely accepted definition of its scope and key functions (Kelly 2006; Sloan 2017; van Efferink 2017). This impacts the outcomes of research. Nonetheless, at its simplest, we can say there are two recognisably distinct schools of geopolitics: classical and critical (Sloan 2017).

1.5 Classical and Critical Geopolitics

Classical geopolitics adopts a modernist epistemological position. That is, there is objective reality in the world which can be investigated empirically by humans, and knowledge of that reality is able to be discovered (whether partially or wholly) by rigorous human inquiry. On this view, one interpretation is possible or correct, and material things have one meaning which is undisputed and consistent across peoples, spaces and times. In this case, this objective reality that might be discovered concerns the structures and patterns through which geography influences the foreign policy decisions, behaviour, and statecraft of nations (Kelly 2016). Kelly adopts this epistemology and an 'objective' search for structures and patterns in defining classical geopolitics as:

> The study of the impact or influence of certain geographic features - these being positions and locations of regions, states, and resources plus topography, climate, distance, immigration, states sizes and shapes, demography, and the like - upon states' foreign policies and actions as an aid to statecraft (Kelly 2016, p. ix).

Of course, as van Efferink notes, it may not always be the case that geography impacts politics—in some instances, it might be the other way round. An alternative view to 'geographic determinism' is that geographic and political processes can have mutual influence (van Efferink 2017).

Sloan offers the following definition of geopolitics, which focuses on the link between geography and the political behavior of states:

> Geopolitical theory is an attempt to draw attention to the importance of certain geographical patterns in political history. It is a theory of spatial relationships and historical causation. From it explanations have been deduced which suggest the contemporary and future political relevance of various geographic conceptualizations (Sloan 1988, cited in Sloan 2017, p. 2).

In stressing that geopolitics is a search for knowledge of spatial relationships and historical causation, Sloan therefore positions geopolitics as a 'synthetic field' of three disciplines: geography, strategic studies and history. Later in this book, as we construct our geopolitical scenarios, we will draw on Sloan's positioning of geopolitics in this way.

In contrast, *critical* geopolitics starts from a post-modern epistemological position that all knowledge (such as geopolitical or geographical knowledge) is inherently subjective, contextual, and contingent. Critical geopolitics seeks to investigate and challenge assumptions in a subjective world in which 'knowledge and power underlie all things' (Kelly 2006, p. 32). This includes questioning the spatial activities of states and the behaviour of actors who influence those activities. It seeks to analyse the artefacts of geopolitics (speeches, texts, maps, narratives, labelling and images) for their role in constructing power over spaces (Wehrmann 2019; Sloan 2017). Critical geopolitics eschews the classical focus on a search for objective knowledge and its emphasis on the links between geography and patterns of behaviour among states. Instead, critical geopolitics attempts to deconstruct or critique existing structures of power and knowledge that seek to spatially order the world. As Sloan (following Deudney) observes, critical geopolitics has a 'strong anti-materialistic vision, one that emphasises the point that all geopolitical constructs serve an ideology' (Sloan 2017, p. 4). It therefore seeks to show how a geopolitical discourse reinforces power relationships. For example, an emerging area of critical enquiry is emotional geopolitics, which considers the affective influence of diplomatic 'performance' to impress and distress, to vocalise and to silence (Jones and Clark 2019). The critical position is normative in that it seeks to uncover and bring to light exploitation of marginalised groups and peoples by those with greater power. As Kelly notes, critical geopolitics advocates for:

> transformation of the current world toward providing more favor to marginal peoples, with an interest in deconstructing foreign policy scripts and other textual materials, confident that such a process will reveal the exploitation being done within existing international hegemony. (Kelly 2006, p. 32)

Thus, where classical geopolitics focuses primarily on patterns and structures of the behaviour of states, critical geopolitics focuses on a plurality of actors, including individual decision-makers, peoples, civil society and private actors. While critical geopolitics should be appreciated for its efforts to make connections across these three vital aspects of understanding Antarctic governance, its post-modern epistemological foundations (Kelly 2006) mean it is a more heterogeneous body of work, often less explicit in its methodology and research design. This heterogeneity is both a strength and a weakness. In particular, it is a challenge for developing a common set of concepts for dialogue between analysts and stakeholders.

This book is about approaches to constructing geopolitical scenarios of Antarctic futures. It is not our purpose to fully explicate or settle the ongoing (but often creative) tension between classical and critical geopolitics. Without wishing to contribute to these tensions, but for the sake of clarity in our work, we proceed with a broad working definition of geopolitics as an investigation of *the intersection between power, authority, and space in rivalry between states*. The key reason we have settled upon this understanding of geopolitics is that, in our view, it has significant advantages for analysis of the Antarctic region which—due to geographic remoteness, difficulties for human activity and the governance arrangements of the ATS—does not have the usual well-defined practices of sovereignty or overt economic and military competition that are present in most other areas of the planet (including Northern polar areas). In Antarctica, rivalry between states is more subtle. We consider our definition of geopolitics well suited to drawing out these subtleties of the Antarctic context. As we explain in Chap. 8, this approach to Antarctic geopolitics might also offer significant value to related fields and disciplines.

In adopting the above definition of geopolitics, we draw on van Efferink's (following Lacoste) conception of geopolitics as 'spatial rivalry',[2] as we feel this is more appropriate to the Antarctic context (at least in the near term) than the stronger term 'conflict'. We propose that *the driving forces of Antarctic geopolitics are likely to be those factors that determine the structures and outcomes of spatial rivalry over the territory, use and resources of Antarctica.* This provides a frame through which we can consider the driving forces (predetermined elements and critical uncertainties) that are central to scenario development.

In adopting a state-centric perspective, we acknowledge that our approach in this book has more affinity with the classical geopolitical tradition. However, we believe this perspective can accommodate the interests both of legal analysts concerned with rule structures, as well as institutionalist analysts concerned with interest-based, cooperative outcomes—see our discussion in Chap. 9. In our view, the 'value-add' of this conception of geopolitics is its focus on the power, authority and spatial rivalry of states, as the principal actors in Antarctic geopolitics. This, we feel, is our central (albeit not exclusive) concern when constructing these scenarios.

1.6 Towards a Critical Realist Geopolitics?

The critiques levelled at both classical and critical schools lead us to consider the possible merits of a 'third way' between these two approaches: a *critical realist* epistemology (Fairclough 2005, cited in Belfrage and Hauf 2017, p. 254). This incorporates 'the most convincing features of both traditions' and accepts that, in addition to the empirical domain of observed experience, there are some processes and events, causal structures and mechanisms that occur whether or not we apprehend them (Belfrage and Hauf 2017). As Tolstoy demonstrates in relation to the contingency of battle, causal powers may be 'possessed unexercised, exercised unrealised, and realised unperceived' (Bhaskar 1998, quoted in Bergene 2007, p. 12).

For this reason, a critical realist perspective acknowledges that a classical, empirically focused approach to Antarctic geopolitics will not yield the full story of the driving forces at work in our scenarios. The critical realist perspective accepts that there are 'brute facts' in the world that are clearly capable of objective knowledge, such as oceans and ice sheets, economic demands and military capabilities. However, the approach also acknowledges that the meaning of such facts is open to contestation by different states, which use subjective interpretations to assert their national interests. For example, maps and textual depictions of the 'Falkland Islands', 'Islas Malvinas', 'Falklands/Malvinas' or 'Malvinas/Falklands' are all alternative framings of the same objective geographic fact. Indeed, without using the critical realist label, Cohen sees geopolitics as the analysis of interactions between political processes,

[2] Leonhardt van Efferink, 'Geopolitical Scenario Planning', Course Notes, Maastricht University Summer School 2019, unpublished.

geographical settings and *perspectives* on those settings (Cohen 1973). It is beyond the scope of this book to develop a critical realist approach to scenario analysis.

We note that a critical realist orientation to geopolitics might complement aspects of scenario practice. As Ramírez and colleagues observe:

> Scenarios are strongly aligned with the critical realist epistemology that holds
>
> that knowledge creation is a social process and cannot be produced in isolation from stakeholders. (Ramírez et al. 2015, p. 84)

At this point, we are unsure whether the underlying assumptions of the classical and critical schools can be reconciled within a single model. However, we suggest in the future this *could be* a pragmatic and fruitful path to explore for narrowing the gap between these two traditions.

1.7 What This Book is About

This book offers one systematic framework to assist policy practitioners, scientists, journalists, commentators, educators, and academics when contemplating plausible futures for Antarctic geopolitics. While there is significant interest in the future of the ATS and Antarctic governance more broadly, much of this work lacks a common language and analytical approach. We suggest scenario planning can provide such a language and approach. Our overarching aim is to explore how to bring the multi-disciplinary focus of geopolitics into Antarctic scenario planning.

The book proceeds as follows. Chapter 2 provides an overview of the history of the ATS and the pervasive presence of geopolitics as a key driver of institutional formation and development in the region. We regard the ATS as an example of effective and stable international governance, but which now faces a new set of geopolitical strains.

Chapter 3 introduces the aims, purposes, and underlying assumptions of scenario planning. We ask: in the context of Antarctic geopolitics, why should we consider scenario planning? And why is the choice of scenario method important? We argue that it is unlikely that most organisations would be prepared to divert time and resources for scenario planning without any expectation that the scenarios created would provide them with a credible picture of what the future might be like. And so, we suggest that it helps to view scenarios as anticipatory systems, with a distinctive type of predictive model, where prediction is not based on probability, but plausibility.

Chapter 4 introduces the key approaches to scenario planning and analytical steps in the scenario planning process. It surveys some of the key options in carrying out a scenario planning process. As a starting point, we adopt the well-known Oxford Scenario Planning Approach (OSPA) to review the deliberate choices we can make in designing scenario planning processes. We introduce and develop the concept of a 'scenario toolbox' and suggest that methods can be grouped loosely under one or more of eight broad approaches. These approaches reflect differences in scenario objectives, the types of data available for use in analysis, and the forms of reasoning that might be employed.

In Chap. 5, we return to geopolitics, and note its concern with the ways that power, authority and spatial (i.e., geographic) factors influence rivalry between states. We see that prominent driving forces of Antarctic geopolitics are likely to be those factors that determine the structures and outcomes of spatial rivalry over the territory, use and resources of Antarctica. We note that the field of Antarctic geopolitics is marked by considerable complexity and uncertainty and, as a result, the driving forces of Antarctic geopolitics exhibit certain dynamics (ways of behaving). We argue that these dynamics need to be understood and captured—at least to some degree-in our scenarios, or they will fail our three-fold test: are they plausible, rigorous and robust?

Chapter 6 provides background to the issue of militarisation (and non-militarisation) in Antarctica and the Southern Ocean, including the question of 'dual- use' equipment and personnel. The chapter reviews the history and present state of military and strategic issues affecting the ATA. This provides the empirical foundation for our case study in Chaps. 7 and 8.

In Chap. 7, we draw upon the scenario methodologies outlined in Chaps. 3–5 to illustrate a systematic approach to constructing scenarios on Antarctic militarisation in the classical geopolitical tradition. For this exercise, we adopt a deductive approach and select the Intuitive Logics 'IL 2 × 2' technique. We augment this technique in several ways to address some of the difficulties encountered when applying it to the field of Antarctic geopolitics.

In Chap. 8, we construct five scenarios on the issue of Antarctic militarisation. Our time horizon is around a decade, and the scenario standpoint we adopt is that of states whose policy and actions preserve and enhance the current rules and norms of the ATS. Here, our purpose is to offer a pragmatic and relevant set of scenarios for stakeholders interested in Antarctic militarisation. From these scenarios, we provide policy recommendations for those many stakeholders interested in preserving the peaceful use of the ATA.

In Chap. 9, we discuss how the scenario analysis on Antarctic militarisation helps us understand the intersection of power, authority and spatial ordering within the ATS. We reflect on how scenarios of Antarctic geopolitics might inform work within more traditional disciplines of Antarctic governance, such as international law and institutional analysis—and may even help us to work across the divide between classical and critical geopolitics.

Chapter 10 concludes our analysis and suggests some paths forward for using scenarios to assess Antarctic futures.

References

Abdel-Motaal, Doaa. 2016. *Antarctica: The Battle for the Seventh Continent*. Santa Barbara: Praeger.

Alonso-Trabanco, Jose Miguel. 2019. Rethinking New Zealand's Grand Strategy. *Geopolitical Monitor*: Backgrounders – October 15. https://www.geopoliticalmonitor.com/rethinking-new-zealands-grand-strategy. Accessed 17 February 2020.

Antarctic Science Advisory Committee (ASAC). 1997. *Australia's Antarctic Program Beyond 2000: A Framework for the Future: A Report to the Parliamentary Secretary for the Antarctic*. Canberra: Department of the Environment.

Belfrage, Claes, and Felix Hauf. 2017. The Gentle Art of Retroduction: Critical Realism, Cultural Political Economy and Critical Grounded Theory. *Organization Studies* 38 (2): 251–271.

Bergene, Anne C. 2007. Towards A Critical Realist Comparative Methodology: Context-Sensitive Theoretical Comparison. *Journal of Critical Realism* 6 (1): 5–27.

Brady, Anne-Marie. 2017. *China as a Polar Great Power*. Cambridge, MA: Cambridge University Press.

Buchanan, Elizabeth. 2021. The Antarctic Treaty System Turns 60. *The Strategist,* 2 July. Canberra: Australian Strategic Policy Institute. https://www.aspistrategist.org.au/the-antarctic-treaty-system-turns-60/. Accessed 29 July 2021.

Buchanan, Elizabeth, and Ryan Burke. 2021. Introducing 'Project 6633' - A Program on Arctic and Antarctic Security. West Point, New York: The Modern War Institute (MWI) at West Point, Department of Military Instruction. https://mwi.usma.edu/introducing-project-6633-a-program-on-arctic-and-antarctic-security/. Accessed 29 July 2021.

Cohen, Saul B. 1973. *Geography and Politics in a World Divided*. New York, NY: Oxford University Press.

Dodds, Klaus, and Mark Nutall. 2016. *The Scramble for the Poles: The Geopolitics of the Arctic and Antarctic*. Cambridge, MA: Polity Press.

Dodds, Klaus, Alan D. Hemmings, and Peder Roberts. 2017. *Handbook on the Politics of Antarctica*. Cheltenham: Edward Elgar.

Future Operating Environment 2035 (Unclassified). 2016. Canberra: Department of Defence. https://www.cove.org.au/wpcontent/uploads/2017/03/Future-Operating-Environment-2035.pdf. Accessed 14 December 2020.

Frame, Bob. 2019. A Typology for Antarctic futures. *The Polar Journal* 9 (1): 236–246.

Frame, Bob. 2020. Towards an Antarctic Scenarios Integrated Framework. *The Polar Journal* 10 (1): 22–51.

Frame, Bob, and Alan D. Hemmings. 2020. Coronavirus at the End of the World: Antarctica Matters. *Social Sciences & Humanities Open* 2 (1): 100054. https://doi.org/10.1016/j.ssaho.2020.100054.

Frame, Bob, Yelena Yermakova, Patrick Flamm, Germana Nicklin, Gabriel De Paula, Renuka Badhe, and Francisco Tuñez. 2021. Antarctica's Gateways and Gatekeepers: Polar Scenarios in a Polarising Anthropocene. *The Anthropocene Review*. https://doi.org/10.1177/20530196211026341.

Guzman, Andrew. 2008. *How International Law Works: A Rational Choice Theory*. Oxford: Oxford University Press.

Haward, Marcus, Julia Jabour, and Brett Peppler. 2003. *Enhancing Australian Influences Within the Antarctic Treaty System in a Turbulent Environment*. Hobart: Antarctic Cooperative Research Centre (CRC).

Hemmings, Alan D. 2014. Re-justifying the Antarctic Treaty System for the 21st Century: Rights, Expectations and Global Equity. In *Polar Geopolitics: Knowledges, Resources and Legal Regimes*, eds. Richard C. Powell and Klaus Dodds, 55–73. Cheltenham, UK: Edward Elgar.

Hemmings, Alan D. 2017. Antarctic Politics in a Transforming Global Geopolitics. In *Handbook on the Politics of Antarctica*, ed. Klaus Dodds, Alan D. Hemmings, and Peder Roberts, 507–522. Cheltenham, UK: Edward Elgar.

Intergovernmental Panel on Climate Change (IPCC). 2019. *IPCC Special Report on the Ocean and Cryosphere in a Changing Climate*. Hans-Otto Pörtner, Debra C. Roberts, Valérie Masson-Delmotte, Panmao Zhai, Melinda Tignor, Elvira Poloczanska, Katja Mintenbeck, et al. eds. https://www.ipcc.ch/srocc. Accessed 10 January 2021.

Jones, Alun, and Julian Clark. 2019. Performance, Emotions, and Diplomacy in the United Nations Assemblage in New York. *Annals of the American Association of Geographers* 109 (4): 1262–1278.

Joyner, Christopher C. 1988. Recommended Measures Under the Antarctic Treaty: Hardening Compliance with Soft International Law. *Michigan Journal of International Law* 19 (2): 401–443.

Joyner, Christopher, and Sudhir Chopta, eds. 1988. *The Antarctic Legal Regime*. Dordrecht: Martinus Nijhoff Publishers.

Kelly, Phil. 2006. A Critique of Critical Geopolitics. *Geopolitics* 11: 24–53.

Kelly, Phil. 2016. *Classical Geopolitics: A New Analytical Model*. Stanford: Stanford University Press.

Kelly, Phil. A Critique of Critical Geopolitics. *Geopolitics* 11(1): 24–53.

Kennicutt, Mahlon C., Steven L. Chown, John J. Cassano, Daniela Liggett, Rob Massom, Lloyd S. Peck, Steve R. Rintoul, et al. 2014a. Polar Research: Six Priorities for Antarctic Science. *Nature* 512 (7512): 23–25.

Kennicutt, Mahlon, Steven L. Chown, John J. Cassano, Daniela Liggett, Lloyd S. Peck, Rob Massom, Steve R. Rinoul, et al. 2014b. A Roadmap for Antarcic and Southern Ocean Science for the Next Two decades and Beyond. *Antarctic Science* 27 (1): 3–18.

Keohane, Robert O., and Joseph S. Nye. 1973. Power and Interdependence. *Survival* 14 (4): 158–165.

Layton, Peter, Travis Hallen, and Lauren Bishop. 2019. *Australia's Antarctic National Air Power Futures*. Canberra: Air Power Development Centre, Department of Defence. https://airpower.airforce.gov.au/publications/australias-antarctica-national-air-power-futures. Accessed 14 December 2020.

Liggett, Daniela, Bob Frame, Neil Gilbert, and Fraser Morgan. 2017. Is it all Going South? Four Future Scenarios for Antarctica. *Polar Record* 53 (5): 459–478.

McGee, Jeffrey, and Marcus Haward. 2019. Australia-China Relations in the Frozen South. *Australian Outlook* - 28 February. Canberra: Australian Institute of International Affairs. https://www.internationalaffairs.org.au/australianoutlook/australia-china-antarctica.

McGee, Jeffrey, and Nengye Liu. 2019. The Challenges for Antarctic Governance in the Early Twenty-first Century. *Australian Journal of Maritime & Ocean Affairs* 11 (2): 73–77.

Natural Environment Research Council (NERC)/British Antarctic Survey (BAS). 2015. The Antarctic Treaty. https://www.bas.ac.uk/about/antarctica/the-antarctic-treaty. Accessed 28 March 2020.

Pettit, Philip. 2019. Social Norms and the Internal Point of View: An Elaboration of Hart's Genealogy of Law. *Oxford Journal of Legal Studies* 39 (2): 229–258.

Pincus, Rebecca, and Saleem H. Ali, eds. 2015. *Diplomacy on Ice: Energy and the Environment in the Arctic and Antarctic*. New Haven: Yale University Press.

Powell, Richard C., and Klaus Dodds, eds. 2014. *Polar Geopolitics: Knowledges, Resources and Legal Regimes*. Cheltenham: Edward Elgar.

Press, Tony, Anne-Marie Brady, Karen Scott, and Julia Jabour. 2019. *Antarctica 2050: Strategic Challenges and Responses*. Canberra: Australian Government, Australian Civil-Military Centre. https://www.acmc.gov.au/resources/publications/antarctica-2050-strategic-challenges-and-responses. Accessed 4 March 2021.

Ramírez, Rafael, Malobi Mukherjee, Simona Vezzoli, and Arnoldo Matus Kramer. 2015. Scenarios as a Scholarly Methodology to Produce 'Interesting Research'. *Futures* 71: 70–87.

Sloan, Geoffrey. 2017. *Geopolitics, Geography and Strategic History*. London: Routledge.

Stokke, Olaf, and Davor Vidas, eds. 1996. *Governing the Antarctic: The Effectiveness and Legitimacy of the Antarctic Treaty System*. Cambridge, MA: Cambridge University Press.

Strating, Bec, and Elizabeth Buchanan. 2021. Why is Antarctica Missing in Action in the Indo-Pacific? 9DASHLINE digital platform. https://www.9dashline.com/article/why-is-antarctica-missing-in-action-in-the-indo-pacific-concept. Accessed 29 July 2021.

Tin, Tina, Daniela Liggett, Patrick T. Maher, and Machiel Lamers, eds. 2014. *Antarctic Futures: Human Engagement with the Antarctic Environment*. Dordrecht: Springer.

Triggs, Gillian. 1986. *International Law and Australian Sovereignty in Antarctica*. Sydney: Legal Books.

Triggs, Gillian, ed. 1987. *The Antarctic Treaty Regime: Law, Environment and Resources*. Cambridge: Cambridge University Press.

van Efferink, Leonhardt. 2017. Geopolitics. In *Oxford Bibliographies*, ed. Barney Warf. New York, NY: Oxford University Press.

Wehrmann, Dorothea. 2019. *Critical Geopolitics of the Polar Regions: An Inter-American Perspective*. Abingdon: Routledge.

Young, Oran R. 2010. *Institutional Dynamics: Emergent Patterns in International Environmental Governance*. Cambridge, MA: The MIT Press.

Young, Oran. R., and Gail Osherenko, eds. 1993. *Polar Politics: Creating International Environmental Regimes*. Ithaca, NY: Cornell University Press.

Chapter 2
History of Antarctic Territorial Claims and Spatial Contestation

Abstract Throughout its history, the ATS has been viewed by many scholars as an example of effective international governance. This is due to its successful responses to internal and external pressures. In particular, for 60 years, the ATS has managed the disputes over sovereignty and security competition that led to its formation. However, the ongoing success of the ATS cannot be taken for granted. We are now in an age of renewed great power competition between China, Russia, and the United States—all key players within the ATS. This new context requires careful management, with skilled leadership and commitment to consensus, as was needed first in the 1950s and then the 1980s and 1990s, when the system was last under significant tension.

Keywords Antarctic territorial claims · Heroic era of Antarctic exploration · Antarctic problem · International geophysical year · Antarctic Treaty · Antarctic Treaty system · Antarctic governance challenges

2.1 Introduction

Given our working definition of geopolitics as *the intersection between power, authority, and space, in rivalry between states*, it is not surprising that the concept resonates through analyses of Antarctic affairs.

Geopolitics has influenced and been at the heart of human endeavours in Antarctica. It was key from the 1890s to the early 1920s, a period termed the 'Heroic Era' of Antarctic exploration (Haward 2017). In late 1893, growing interest in Antarctica saw two lectures delivered to the Royal Geographical Society in London. On 13 November, Sir Clements Markham, President of the Society, delivered a lecture titled 'Geographical Desiderata, or Exploring Work to be Done and Geographical Problems to be solved'; a fortnight later, Dr John Murray spoke on 'The Antarctic Region and the Science and Commercial Results of its Exploration' (Royal Geographical Society 1893). Murray urged more research in the Antarctic to 'resolve the outstanding geographical questions still posed in the south' (Royal Geographical Society 1893). While discovery and science were motives for exploring Antarctica in the 1890s, territorial acquisition, or at least the planting of flags and claiming territory, were also key acts of the 'Heroic Era'. In 1912, Roald Amundsen's success in the 'race to

J. McGee et al., *The Future of Antarctica*, Springer Polar Sciences,
https://doi.org/10.1007/978-981-16-7095-4_2

the South Pole' against Robert Falcon Scott's team—in many ways overshadowed by the death of Scott and his companions (Huntford 2012)—included Amundsen claiming the polar plateau in the name of King Haakon VII of Norway.

2.2 The 'Antarctic Problem'

The intersection of power, authority, and space, through territorial claims and assertions of national sovereignty, were the critical elements of the 'Antarctic Problem' (Christie 1951) which arose from the seven territorial claims to Antarctic made between 1908 and 1942 by the United Kingdom, New Zealand, France, Norway, Australia, Chile and Argentina (Wilson 1964) (see Table 2.1 and Fig. 2.1). It would become a key geopolitical conundrum in the mid-twentieth century: Chile's 1940 claim to Antarctic territory, and Argentina's 1942 claim to Antarctic territory, as amended in 1946 (Wilson 1964), overlapped the UK's 1908 claim. While Chile formally announced its territorial claim in 1940, Oscar Pinochet De La Barra suggests that it had 'perfected [its] title in Antarctica' in 1906 (De La Barra 1955). The UK's *Operation Tabarin* in 1943—ostensibly a meteorological expedition to the Antarctic Peninsula—was clearly a response to Chilean and Argentinian actions in asserting territorial sovereignty (Dudeney and Walton 2012).

Following the United Kingdom's claim to the British Antarctic Territory and its transfer of the Ross Dependency to New Zealand in 1923, France announced its territorial claim on 29 March 1924. This announcement did not include 'precise statements of the limits of this region' (Miller 1927, p. 508). Agreement over the

Table 2.1 Antarctic territorial claims and dates of claim

Year of claim	Claimant
1908	British Antarctic Territory
1923[a]	Ross Dependency (New Zealand)
1924	Adélie Land (France)
1929	Peter I Island (Norway)
1933[b]	Australian Antarctic Territory
1939	Queen Maud Land (Norway)
1940	Chilean Antarctic Territory
1942[c]	Argentine Antarctica

[a] Transferred to New Zealand—following original claim by James Clark Ross in 1941

[b] Australia completed the transfer of the Australian Antarctic Territory from the United Kingdom in 1936

[c] Argentina's claim was amended in 1946 with the Western limit of Argentina's territory at the 74th meridian (Huntford 2012, p. 23). The Western boundary had previously been claimed to be Longitude 68° 34′ W

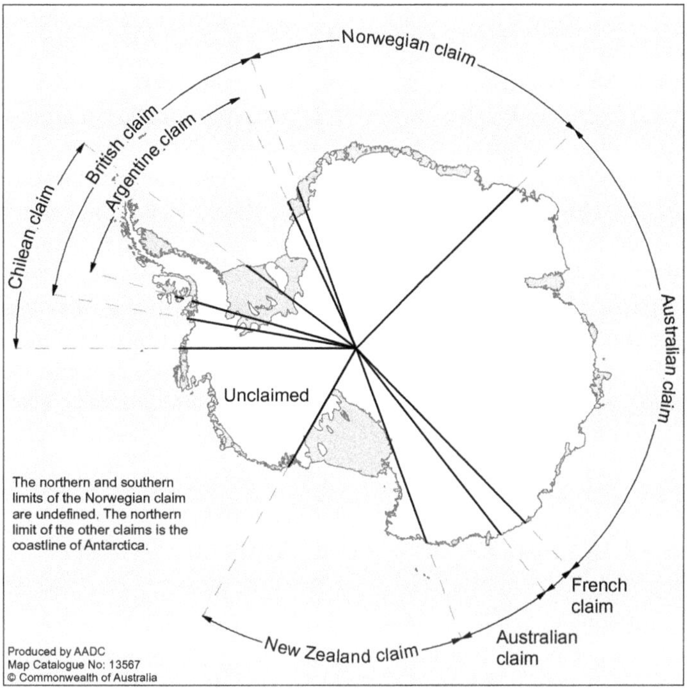

Fig. 2.1 Antarctic territorial claims. *Source* AAD Data Centre. https://data.aad.gov.au/database/mapcat/antarctica/territorial_claims_b&w_13567.pdf

boundaries for the French claim was finally reached in March 1938 (Miller 1927). The Imperial Conference of 1926 [an intergovernmental meeting of delegations from the governments of the UK, Canada, the Commonwealth of Australia, New Zealand, the Union of South Africa, the Irish Free State, Newfoundland, and India (National Library of New Zealand 1927)] resolved that the entire Antarctic continent should become a possession of Britain and its dominions, New Zealand, and Australia. The members of the conference agreed on the way this could be achieved (National Archives of Australia 1926).

Other states also had potential territorial interests in the Antarctic. In the nineteenth century, the United States carried out pioneering sea voyages to the deep reaches of the Southern Ocean, led by Palmer in 1820 and Wilkes in 1840 (Joyner 1982). From the late 1920s to 1940, Admiral Richard E. Byrd led three voyages to Antarctica and established Little America station on the Ross Ice Shelf (Hall 1989). Byrd's Third Expedition (1939–41) was to establish permanent stations to support a US claim to territory within the framework of the Hughes doctrine[1] (Wilson 1964), where the US viewed discovery, even when followed by a formal act of possession, as not

[1] The Hughes doctrine is named after Charles E. Hughes, United States Secretary of State, who in 1924 made statements that the United States was reluctant to make a claim to Wilkes Land in Antarctica.

supporting a valid claim to territory unless it was followed by actual settlement (Hall 1989).

While Chile and Argentina made claims during the years of the Second Word War, Japan and Germany also had potential claims to parts of the Antarctic continent. The Japanese Antarctic Expedition of 1910–12, led by Nobu Shirase in the *Kainan Maru*, saw the first non-European attempt to reach the South Pole (Stevenson 2010). Though unsuccessful in his attempt to reach the Pole, Shirase did raise the Rising Sun flag over the Yamamoto Snowfield (Latitude 80° 05′ S and Longitude 156° 37′ E) (Hall 1989). In 1938, Nazi Germany sent the ship *Schwabenland* to Antarctica (Anon. 1940) to explore areas between the United Kingdom and Norwegian claims (between longitude 20 E and 10 W) referred to as *Neu-Schwabenland* (Lüdecke 2012). The German expedition was also seeking new whaling grounds, as whale oil remained a key strategic commodity in the years leading up to the War (Lüdecke 2012).

2.3 Geopolitics and Resolving the 'Antarctic Problem'

Chile and Argentina's actions to assert and formalise territorial claims during the years of the Second World War meant that the 'Antarctic Problem' remained very much alive in international politics. In the post-war period, the 'Antarctic Problem'—along with other, immediate geopolitical 'problems'—remained to be resolved (Christie 1951), as increasing geopolitical tension and great power rivalry gave way to the Cold War. In 1947–48, the United States, having made no territorial claim itself, proposed that Antarctica be governed under a United Nations trusteeship arrangement. Britain instead proposed the establishment of an eight-power 'condominium' or joint sovereignty by all seven of the claimant states—plus the US—for Antarctica. Argentina and Australia did not support either proposal. Norway viewed them as unnecessary, while France requested more information.

Chile took the position that a trusteeship of Antarctica would not be possible under the UN Charter (Jacobsson 2011), and instead proposed a joint declaration of all interested states, effectively freezing all claims over Antarctic territory for a period of 5–10 years (Jacobsson 2011). This proposal has been termed the 'standstill agreement' or the 'Escudero approach' after the Chilean diplomat, academic and international lawyer, Professor Julio Escudero, who suggested this initiative to resolve tensions. In the first sign of demilitarisation of the area, in late 1948 the United Kingdom agreed with Argentina that, to ease territorial tensions in the Antarctic peninsula, it would not send naval vessels south of 60 degrees South in the following season (United States Department of State 1948).

In 1949, the Union of Soviet Socialist Republics (the Soviet Union) indicated (and, in 1950, formally announced) that any decision about Antarctica made without its participation would be, in its view, illegitimate. The Soviet Union's possible territorial interests in Antarctica are linked back to activities sponsored by imperial Russia through Captain Bellingshausen's naval expedition to the deep Southern Ocean.

[Bellingshausen's 1819–21 voyage with the ships *Vostock* and *Mirny* is credited as the first sighting and the first circumnavigation of the continent (United States Department of State 1948, p. 835).] The Soviet Union's interest in Antarctica was a focus of concern for the United States government, with differences in strategy evident between the Department of Defense, which wanted to exclude the Soviet Union, and the Department of State, which recommended ongoing engagement (Jacobsson 2011).

In 1949, further developments occurred. The United States and the United Kingdom discussed a new proposal, based on Chile's 'standstill agreement', to place a time-limited freeze on Antarctic territorial claims. The UK responded positively to this proposal and the US initiated informal discussions based on the proposal with Chile, Australia and New Zealand. The outbreak of the Korean War (1950–53) saw the discussions stall, and tensions over territorial claims in the Antarctic Peninsula again emerged between Chile, Argentina, and the UK in 1952–53. This included an incident at Hope Bay in February 1952, when the first and only hostile gunshots were fired in Antarctica by Argentinian troops to ward off a British research vessel (Boczek 1984).

At the same time, however, collaborative science activities were emerging between key states involved in these discussions over the future of Antarctic as precursors to the increased international scientific efforts during the International Geophysical Year of 1957–58. In 1956 and 1957, some US scientists overwintered at the Soviet Union's Mirny research station in East Antarctica, while Soviet scientists worked at the Little America station on the Ross Ice Shelf. Australian scientists also visited the Mirny station during its construction in the austral summer of 1956 as part of an Australian National Antarctic Research Expedition (Beck 2014).

This collaboration was not free from global or local geopolitical tensions, however. A Soviet proposal for a port visit to the Australian port of Adelaide on the return voyage of the Soviet expedition prompted extensive domestic political debate—at a time of considerable tension in Australia over the influence of the Soviet Union and local communists in Australian domestic affairs (Bowden 2011). Intervention by the then-Australian Foreign Minister, Sir Richard Casey, enabled the Soviet vessel to visit Adelaide, while expeditioners were able to meet with Sir Douglas Mawson, who then held an academic position at the University of Adelaide (Bowden 2011).

Developing world interest in Antarctica was also noteworthy during the 1950s. India was an important figure in the non-aligned movement after the Second World War and played a major role in attempts to bring the Antarctic under United Nations control through a trusteeship arrangement. Indeed, the discussion over Antarctica raised by India and Sweden in 1956 (Elzinga 2013) foreshadowed later actions by Malaysia in the United Nations General Assembly in the 1980s. India's 1956 initiative did not, however, proceed, being overtaken by other Cold War concerns at the United Nations (Bowden 2011). In the 1970s, India re-engaged with Antarctica, developed significant scientific programs and set aside its earlier concerns over the Antarctic Treaty. India acceded to the Antarctic Treaty in August 1983, gaining Consultative Party status within a month of accession (Chaturvedi 2013).

Escudero's 'standstill agreement', promoted by the United States in 1949, provided a way forward in Antarctic diplomacy. Scientific collaboration in Antarctica, together with the 'standstill' concept on territorial claims, was at the core of the ideas that supported interest in the Antarctic component of the International Geophysical Year (IGY) of 1957–58. The IGY marked the beginning of the modern era of scientific research in Antarctica and was a key event in moving to the eventual solution of the 'Antarctic Problem'. The IGY had been initiated through discussions among leading scientists, led by the US' Professor James Van Allen in 1950 (Stoddart and Haward 2011), which followed the rapid development of science and technology in the Second World War and immediate post-war period. The IGY addressed the need for improved scientific research in Antarctica, with a focus on the disciplines of meteorology, geophysics, glaciology, oceanography, and seismology.

For the first time, Antarctica was made a focus for concerted and collaborative research. For this purpose, a number of purpose-built national Antarctic research stations were opened during the IGY, including the UK's Halley station, Japan's Swoya station, and several Soviet stations, such as Vostok station, high on the ice sheet in the Australian- claimed sector of East Antarctica (Stoddart and Haward 2011). (Australia's Mawson station, constructed in 1954, predated the IGY by three years.) States taking part in the IGY could not use activities under the program as a basis to make or expand territorial claims. Similarly, states with existing territorial claims could not use IGY activities to bolster their own claims. Nonetheless, geopolitics, as seen through spatial contestation, remained a significant influence on state behaviour in Antarctica.

The South Pole station (Amundsen-Scott base) was established by the United States in a deliberate attempt to forestall the Soviet Union placing a station at the geographic South Pole and thereby potentially dominating the continent. In response, the Soviet Union established research stations in all the claimant states' territories in the period 1955–58. After the Soviet Union constructed Vostock station in 1956, Australia responded diplomatically, *post-facto*, by inviting the Soviets to do so (Boczek 1984). The invitation was effectively ignored, as the Soviet Union did not recognise the Australian claim in East Antarctica.

The IGY did, however, diffuse tensions between key Antarctic states and saw collaboration and sharing of logistics, even between Cold War foes, the United States, and the Soviet Union. Links between Soviet and Australian scientists also continued to develop after the IGY, despite Cold War geopolitical rivalry between the West and the Eastern bloc (Elzinga 2013). The IGY is rightly seen as a major factor in the development of the Antarctic Treaty. Yet, the argument that science collaboration was the prime motive (Budd 2002) ignores the broader geopolitics of the period. The success of science in the IGY was a necessary, but not a sufficient, condition for the conclusion of signing of the Antarctic Treaty in December 1959 in Washington, D. C.

2.4 The Antarctic Treaty: Setting Aside Geopolitical Differences?

Negotiation of the Antarctic Treaty took eighteen months between June 1958 and December 1959. 60 preparatory meetings were held between June 1958 and October 1959, with a full diplomatic conference beginning on 15 October 1959 and concluding with the signing of the Treaty on 1 December 1959 (Hanessian 1960). This was a significant diplomatic effort, led by the United States, to balance the aspirations and interests of a number of different states (Haward 2017). The 1959 Antarctic Treaty was concluded and entered into force during this period of heightened Cold War great power competition between the US and the Soviet Union (Haward and Griffiths 2011). The Antarctic Treaty effectively managed this competition by demilitarising the Antarctic continent and directing all human presence on the continent towards peaceful use and scientific research.

The Antarctic Treaty also furthered Escudero's 'standstill approach' by setting aside disputes over territorial claims by the provisions of Article IV (Scott 2011). Article IV effectively put into treaty language the practice of the standstill proposal as utilised in the IGY. Article IV is seminal to the spatial ordering of the Antarctic continent over the last 60 years, so worth recounting in full:

> 1. Nothing contained in the present treaty shall be interpreted as:
>
> (a) a renunciation by any Contracting Party of previously asserted rights of or claims to territorial sovereignty in Antarctica;
>
> (b) a renunciation or diminution by any Contracting Party of any basis of claim to territorial sovereignty in Antarctica which it may have whether as a result of its activities or those of its nationals in Antarctica, or otherwise;
>
> (c) prejudicing the position of any Contracting Party as regards its recognition or non-recognition of any other States right of or claim or basis of claim to territorial sovereignty in Antarctica.
>
> 2. No acts or activities taking place while the present treaty is in force shall constitute a basis for asserting, supporting or denying a claim to territorial sovereignty in Antarctica or create any rights of sovereignty in Antarctica. No new claim, or enlargement of an existing claim, to territorial sovereignty in Antarctica shall be asserted while the present treaty is in force (The Antarctic Treaty 1959).

The position of the Antarctic Treaty parties regarding territorial claims (whether existing or potential) are 'put on hold' as at the year 1961, but not abandoned. The United States and the Soviet Union made reservations over their 'claims to any or all of Antarctica' on lodging their instruments of ratification to the Antarctic Treaty. Importantly, this Treaty is an agreement to use the continent for peaceful purposes and to forego the practical assertion of territorial claims whilst the Treaty remains in force. The Antarctic Treaty has no end date, so, unless the parties decide to terminate the Treaty, it continues in perpetuity.

In summary, the Antarctic Treaty:

- stipulates that Antarctica should be used exclusively for peaceful purposes—military activities, such as the establishment of military bases or weapons testing, are specifically prohibited (Art I);
- guarantees freedom to conduct scientific research (Art II);
- promotes international scientific cooperation and requires that the results of research be made freely available (Art III);
- sets aside the potential for sovereignty disputes between Treaty parties (Art IV);
- prohibits nuclear explosions and the disposal of radioactive waste (Art V);
- defines the geographic operation of the Treaty as the area south of 60 degrees South Latitude (Art VI)
- provides for inspection to ensure compliance with the Treaty—a world-first inspection system (Art VII);
- requires advance notice of expeditions (Art VII);
- provides for the exercise of domestic legal jurisdiction over observers and scientific staff on exchange (Art VIII); and
- provides for the parties to discuss measures to further the Treaty (Art IX) (Bergin and Haward 2007).

Article VI of the Antarctic Treaty defines its area of application as south of 60 degrees latitude, including ice shelves. While the Treaty applies to all marine areas below 60 degrees South, it explicitly states that it does not prejudice high seas rights in the region (e.g., fishing, navigation, undersea cables) in international law of the sea. The harvesting of marine mammals under such high seas rights are specifically governed by other treaties (ICRW and CCAS), so not within the jurisdiction of the Antarctic Treaty. The treatment of airspace is considered by analysts to follow the status of the underlying surface (Oduntan 2012, p. 145).

Over the last 50 years, international governance of Antarctica has steadily expanded in scope. In 1959, during the geopolitical tension of the early Cold War period, the Antarctic Treaty was formed by the 12 original states primarily to prevent security competition in the Antarctic region (Haward and Griffiths 2011). Since that time, international governance of Antarctica and the Southern Ocean has steadily expanded through the formation of several issue-specific treaties complementing the Antarctic Treaty: the 1972 Convention for the Conservation of Antarctic Seals; the 1982 Convention on the Conservation of Antarctic Marine Living Resources (CAMLR Convention); and the 1991 Protocol on Environmental Protection to the Antarctic Treaty (Madrid Protocol). In 2001, a related treaty, the Agreement on the Conservation of Albatrosses and Petrels (ACAP), was formed to provide protection to sub-Antarctic seabirds. Collectively known as the Antarctic Treaty System (ATS), this related group of treaties and the binding measures of the yearly meeting of parties to the Antarctic Treaty—the ATCM—are treated in the academic literature as a leading example of effective international environmental governance (Joyner 1998).

Resolution—or at least management—of geopolitical tension been a constant in the establishment and operation of the Antarctic Treaty System (Budd 2002). During

Table 2.2 Number of parties to the Antarctic Treaty by decade of accession

Period	Antarctic Treaty Consultative Parties (ATCPs)	Non-consultative parties (NCPs)
1961–70	12	1
1971–80	1	0
1981–90	13	11
1991–00	1	4
2001–10	1	4
2011–18	1	5
Total	29	25

Source Antarctic Treaty Secretariat

the period from the 1970s to the 1990s, the ATS demonstrated significant ability to respond and adapt to major geopolitical tensions and challenges (Bergin and Haward 2007, p. 56) that, in turn, were directed by external challenges and internal debates within the ATS. The developing world's 'Question of Antarctica' from 1982–2002, discussed further below, posed a major challenge to the ATS and its governance of the region. In the late 1980s, debate over the proposed Convention on the Regulation of Antarctic Mineral Resources Activities (CRAMRA) posed a moment of potential crisis, a breakdown in consensus decision-making, within the ATS.

The period from 1981 to 1990 was also a time of significant expansion in the number of Antarctic Treaty Consultative Parties (ATCPs) which have decision-making rights within the annual meeting of the Antarctic Treaty (see Table 2.2). The number of new entrants to the Antarctic Treaty in this decade represented a marked change from the first two decades after the entry into force of the Treaty, when the seven claimant states formed a majority of the original 12 (later 13) ATCPs.

Membership of the Antarctic Treaty has significantly expanded from its original 12 to 54 states. India achieved ATCP status in 1983; China followed in 1985, two years after acceding to the Treaty. More recently, acceding members as NCPs include Monaco (2008), Malaysia (2011), Pakistan (2012), Mongolia (2015), Iceland (2015) and Kazakhstan (2015).

In 1983, Malaysia raised its concern over the governance of Antarctica through its leadership of the 'Question of Antarctica' debate. Malaysia's significant and sustained critique of the ATS led to debate in the United Nations General Assembly (UNGA) on the 'Question of Antarctica' from 1983. Then-Prime Minster Dr Mahathir questioned the operation of the ATCM and the alleged failure of the ATS to represent aspirations of developing states. He also critiqued the ATS as a neo-colonial closed shop (Bowden 2011; Haward and Mason 2011), and pointed to the need to reflect the 'common heritage of mankind' principle in Antarctica. The 'Question of Antarctica' became a matter for the UNGA over a 20-year period between 1982–2002, without undermining the Antarctic Treaty (Tepper and Haward 2005). Eventually, Malaysia—the leading country in this move within the UNGA—acceded to the Antarctic Treaty on 31 October 2011 (Haward and Mason 2011). It continues to develop its science program and is now a full member of the Scientific Committee on Antarctic Research (SCAR).

The 'common heritage of mankind' does not appear in the Antarctic Treaty, but in the 1979 UN Moon Treaty for lunar resources and 1982 UN Convention on the Law of the Sea for deep seabed mining. The 'common heritage of mankind' principle provides for a compulsory sharing of the benefit of extractive activity between developed and developing states. Prime Minister Mahathir used Malaysia's chairing of the non-aligned movement to promote this issue and critique the ATS (Jackson and Boyce 2011). The geopolitical focus increased in salience as Dr Mahathir engaged with African 'Front Line States' and linked the issue to South Africa's participation in the ATS (Joyner 1998).

While the ATCPs as a group stood firm in opposing Malaysia's proposal for greater UN involvement in Antarctica, individual ATCPs found it more difficult to 'hold the line'. The Soviet Union, for example, was challenged in supporting the ATCP position when Malaysia increasingly linked South Africa's role as an ATCP to the debate. Malaysia sought the support of African states opposed to South Africa's policy of apartheid (Haward and Mason 2011). The response to Malaysia in the UNGA debate was led by Australia, whose position in this was strengthened by the wide experience of its representative to the UN, senior diplomat Richard Woolcott. Woolcott's previous diplomatic postings included Malaysia, South Africa, Ghana, and the Soviet Union (Woolcott 2003), and his personal connections with senior government figures in the developing world played a major role in managing this issue (Woolcott 2003). Woolcott's responses focused on the strength of the ATS. He also worked hard to ensure Malaysia increased its awareness of the workings of the ATS and understood the position of the ATCPs (Haward and Mason 2011).

At the same time, debate outside the ATS increasingly concerned the adequacy of environmental protection. This was particularly evident in relation to proposals to regulate mining in Antarctica through the mid- to late-1980s negotiation of the Convention for the Regulation of Antarctic Mineral Resource Activities (CRAMRA). Environmental non-government organisations in particular increased their attention on Antarctica as a result. For example, Greenpeace established and operated its 'World Park' Antarctica Base between 1987–1991 at Cape Evans in the Ross Sea region. Following the successful negotiation of the 1991 Protocol on Environmental Protection to the Antarctic Treaty (Madrid Protocol), the Greenpeace base was dismantled and removed in 1992.

Australia's 1989 decision not to ratify CRAMRA, despite having agreed to the text, was very significant, and was criticised as destabilising the ATS. The Australian Government's opposition to signing CRAMRA, led strongly by then-Prime Minister Robert (Bob) Hawke, centered on the issues of environmental protection of Antarctica, safeguarding of Australia's territorial claim, and concerns over the subsidisation of mineral exploration and extraction (which might negatively impact upon export markets for Australian mining) (Woolcott 2003). France—urged on by the charismatic advocacy of famous marine scientist, Jacques Cousteau—was seeking to rebuild its environmental reputation, following damage caused during attempts to develop an airstrip at its Dumont D'Urville station, as well as the ongoing fallout from the 1985 bombing of Greenpeace's *Rainbow Warrior* in Auckland Harbour, and its Pacific nuclear testing program. French Prime Minister Michel Rocard also bonded strongly with the Australian Prime Minster to lead opposition to CRAMRA (Jackson and Boyce 2011).

Australia and France therefore opened up a debate among ATCPs that was pivotal in setting aside this instrument and in the negotiations that led to the development of the 1991 Protocol on Environmental Protection to the Antarctic Treaty (Madrid Protocol). France and Australia, as the two major protagonists leading the charge against ratifying CRAMRA, were joined by other ATCPs and gained support from an increasing array of non-governmental groups. New Zealand, the United Kingdom and the United States opposed the Australia-France initiative at the ATCM, but political opinion at the highest levels in the UK and US was shifting away from ratifying CRAMRA—and moving faster than the positions held by their key Antarctic officials.

The 1991 Madrid Protocol entered into force in 1998 for the text and Annexes I–IV. Annex V on Area Protection and Management, being subject to a slightly different ratification process, was not concluded at the Madrid Special ATCM, so entered into force later, on 24 May 2002. Annex VI on Liability Arising from Environmental Emergencies was finalised in June 2005 but is yet to enter into force. Article 2 of the Madrid Protocol states that, 'The Parties commit themselves to the comprehensive protection of the Antarctic environment and dependent and associated ecosystems and hereby designate Antarctica as a natural reserve, devoted to peace and science' (Protocol on Environmental Protection to the Antarctic Treaty 1991). The Madrid Protocol is thus highly significant in adding environmental protection to 'peace' and 'science' as one of the core norms of the ATS.

2.5 Geopolitics and the Future of Antarctica

Throughout its history, the ATS has been viewed by many institutionalist and international legal scholars as an example of effective international governance (Young 2010) because of its successful responses to internal and external pressures (Young 2010, p. 530). However, the ATS now potentially faces a new set of geopolitical strains. Membership of the Antarctic Treaty has significantly expanded from its original 12 states; the 54 current states in the Antarctic Treaty include China and India, as well as newly acceding members, such as Malaysia, Pakistan, Iran and Iceland. The ATS also faces significant governance challenges from rapidly increasing tourism (particularly from cruise ships operating out of Chile and Argentina), fishing and bioprospecting—all economic uses that create new interests and potential geopolitical tensions. A warming climate is a key driver of change within Antarctic and Southern Ocean environments (Constable et al. 2014). This change will likely generate new geopolitical tensions between states, including tensions arising from range shifting of commercially valuable marine species from mid to high latitudes

(Hobday and Pecl 2014). There also remains ongoing discussion of the mining prohibition under the Madrid Protocol (Vöneky et al. 2012; Press 2015).

The world also appears to be moving into a new period of great power competition between China, Russia and the United States—all key players within the ATS. Looking back, the Antarctic Treaty has been successful in ensuring that geopolitical tensions have been minimised. Antarctica and the Southern Ocean have remained free from military conflict, retained their non-militarised status, and decision-makers within the ATCM meetings have arguably struck a balance between the interests of both claimant and non-claimant states. In this sense, the system of governance that has developed from the Antarctic Treaty has been effective in a 'problem solving' sense; that is, it has avoided the dispute over sovereignty and security competition that led to its formation (Haward and Griffiths 2011; Stokke and Vidas 1996; Young 2014).

The ongoing success of the ATS cannot, however, be taken for granted, particularly in an age of renewed great power competition. This new context will likely require careful management, with skilled leadership and commitment to consensus, as was needed first in the 1950s and then the 1980s and 1990s, when the system was last under significant tension. This volume hopes to contribute to this project by examining how geopolitical scenario analysis might be used by those interested in Antarctic governance to better understand plausible geopolitical futures for the region and to help build capacity within the stakeholder community for appropriate policy responses.

References

Anon. 1940. 'The *Schwabenland* in the Antarctic. *The Geographical Journal* 95 (1): 52–54.

Beck, Peter. J. 2014 (1986). *The International Politics of Antarctica.* Abingdon, UK: Routledge Revivals.

Bergin, Anthony, and Marcus Haward. 2007. *Frozen Assets: Securing Australia's Antarctic Future.* Strategic Insight No 34. Canberra: Australian Strategic Policy Institute.

Boczek, Boleslaw A. 1984. The Soviet Union and the Antarctic Regime. *The American Journal of International Law* 78 (4): 834–858.

Bowden, Tim. 2011. Phillip Garth Law: The Architect of ANARE. In *Australia and the Antarctic Treaty System: Fifty Years of Influence*, ed. Marcus Haward and Tom Griffiths, 140–146. Sydney: UNSW Press.

Budd, W. F. 2002. Scientific Imperative for Antarctic Research. In *The Antarctic: Past, Present and Future*, eds. Julia Jabour-Green and Marcus Haward, 41–59. Hobart: Antarctic CRC Research Report #28.

Chaturvedi, Sanjay. 2013. Antarctica as Global Knowledge Commons: Geopolitics, Science and Trusteeship. In *Scientific and Geopolitical Interests in Arctic and Antarctica*, eds. R. Ramesh, M. Sudhakar, and Sulagna Chattopadhyay, 11–32. New Delhi: LIGHTS Research Foundation.

Christie, E. W. Hunter. 1951. *The Antarctic Problem: An Historical and Political Study.* London: George Allen and Unwin.

Constable, Andrew J., Jessica Melbourne-Thomas, Stuart P. Corney, Kevin R. Arrigo, Christophe Barbraud, David K. A. Barnes, Nathaniel L. Bindoff, et al. 2014. Climate change and Southern

Ocean Ecosystems I: How Changes in Physical Habitats Directly Affect Marine Biota. *Global Change Biology* 20 (10): 3004–3025.
De La Barra, Oscar Pinochet. 1955. *Chilean Sovereignty in Antarctica*, Santiago, Chile: Editorial Del Pacifico.
Dudeney, John R., and David Walton. 2012. From Scotia to 'Operation Tabarin': Developing British Policy for Antarctica. *Polar Record* 48 (4): 342–360.
Elzinga, Aant. 2013. Rallying Around a Flag? On the Persistent Gap in scientific internationalism in Deed and Word. In *The Emerging Politics of Antarctica*, ed. Anne-Marie. Brady, 193–213. Abingdon, UK: Routledge.
Hall, H. Robert. 1989. The 'Open Door' into Antarctica: An Explanation of the Hughes Doctrine. *Polar Record* 25 (153): 137–140.
Hanessian, John. 1960. The Antarctic Treaty 1959. *The International and Comparative Law Quarterly* 9 (3): 436–480.
Haward, Marcus. 2017. The Originals. In *Handbook on the Politics of Antarctica*, ed. Klaus Dodds, Alan D. Hemmings, and Peder Roberts, 232–240. Cheltenham, UK: Edward Elgar.
Haward, Marcus, and David Mason. 2011. Australia, the United Nations and the Question of Antarctica. In *Australia and the Antarctic Treaty System: Fifty Years of Influence*, ed. Marcus Haward and Tom Griffiths, 202–221. Sydney: UNSW Press.
Haward, Marcus, and Tom Griffiths, eds. 2011. *Australia and the Antarctic Treaty System: Fifty Years of Influence*. Sydney: UNSW Press.
Hobday, Alistair. J., and Gretta T. Pecl. 2014. Identification of Global Marine Hotspots: Sentinels for Change and Vanguards for Adaptation Action. *Reviews in Fish Biology and Fisheries* 24(2): 415–425.
Huntford, Roland. 2012. *Race for the South Pole: The Expedition Diaries of Scott and Amundsen*. London: Bloomsbury.
Jackson, Andrew, and Peter Boyce. 2011. Mining and 'World Park Antarctica, 1982–1991. In *Australia and the Antarctic Treaty System: Fifty Years of Influence*, ed. Marcus Haward and Tom Griffiths, 243–273. Sydney: UNSW Press.
Jacobsson, Marie. 2011. Building the international legal framework for Antarctica. In *Science Diplomacy: Antarctica, Science, and the Governance of International Spaces*, eds. Paul A. Berkman, Michael Lang, David W. H. Walton, and Oran R. Young, 1–15. Washington D.C.: Smithsonian Institution Scholarly Press.
Joyner, Christopher C. 1982. *The Law of the Sea and Antarctica*. Dordrecht: Martinus Nijhoff Publishers.
Joyner, Christopher C. 1998. Recommended Measures Under the Antarctic Treaty: Hardening Compliance with Soft International Law. *Michigan Journal of International Law* 19 (2): 401–443.
Lüdecke, Cornelia. 2012. Investigation of the Unknown: The Flight Programme of the German *Schwabenland* Expedition 1938/39. *The Polar Journal* 2 (2): 231–333.
Miller, David Hunter. 1927. National Rights in the Antarctic. *Foreign Affairs* 5 (3): 508–510.
National Archives of Australia. 1926. *Report of the Committee on British Policy in the Antarctic to the Imperial Conference*. NAA: A981, ANT 4 PART 4.
National Library of New Zealand. 1927. *Imperial Conference 1926: Summary of Proceedings. Appendix to the Journals of the House of Representatives, 1927 Session I, A-06*. New Zealand: AtoJs Online. https://atojs.natlib.govt.nz/cgi-bin/atojs?a=d&d=AJHR1927-I.2.1.2.9.
Oduntan, Gbenga. 2012. *Sovereignty and Jurisdiction in the Airspace and Outer Space*. Abingdon, UK: Routledge.
Press, Anthony J. 2015. The Antarctic Treaty System: Future Mining Faces Many Mathematical Challenges. *The Yearbook of Polar Law* 7 (1): 623–631.
Protocol on Environmental Protection to the Antarctic Treaty, opened for signature 4 October 1991, 30 ILM 1455 (entered into force 14 January 1998).
Royal Geographical Society. 1893. Session 1893–4. *The Geographical Journal* 2 (5): 1–4.
Scott, Shirley V. 2011. Ingenious and Innocuous? Article IV of the Antarctic Treaty as Imperialism. *The Polar Journal* 1 (1): 51–62.

Stevenson, William R. 2010. Science, the South Pole and the Japanese Expedition of 1910–1912. *Endeavour* 35 (4): 160–168.
Stoddart, Michael, and Marcus Haward. 2011. Science. In *Australia and the Antarctic Treaty System: Fifty Years of Influence*, ed. Marcus Haward and Tom Griffiths, 138–163. Sydney: UNSW Press.
Stokke, Olaf, and Davor Vidas, eds. 1996. *Governing the Antarctic: The Effectiveness and Legitimacy of the Antarctic Treaty System*. Cambridge: Cambridge University Press.
Tepper, Rohan, and Marcus Haward. 2005. The Development of Malaysia's Position on Antarctica: 1982–2004. *Polar Record* 41 (2): 113–124.
The Antarctic Treaty, opened for signature 1 December 1959, 402 UNTS 71 (entered into force 23 June 1961).
United States Department of State. 1948. *Foreign Relations of the United States*, 1948, General; The United Nations, Vol 1 Part 2. Document 260 800.014 Antarctic/11–2448: Telegram: 1015. https://history.state.gov/historicaldocuments/frus1948v01p2/d260. Accessed April 14, 2020.
Vöneky, Silja, and Sange Addison-Agyei. 2012. Antarctica. In *Max Planck Encyclopedia of Public International Law*, vol. 1, ed. Rüdiger. Wolfrum, 418–435. Oxford: Oxford University Press.
Wilson, Robert E. 1964. National Interests and Claims in the Antarctic. *Arctic* 17 (1): 15–31.
Woolcott, Richard. 2003. *The Hot Seat: Reflections on Diplomacy from Stalin's Death to the Bali Bombings*. Sydney: Harper Collins.
Young, Oran. R. 2014. Building an International Regime Complex for the Arctic: Current status and next steps. *The Polar Journal* 2(2): 391–407.
Young, Oran. R. 2010. *Institutional Dynamics: Emergent Patterns in International Environmental Governance*. Cambridge, MA: The MIT Press.

Chapter 3
Scenario Analysis and the Limits of Prediction

Abstract The limits of prediction pose a critical risk for any organisation making long-range decisions about its strategy and investments. The accuracy of our forecasts degrades progressively due to the combined effects of complexity and uncertainty, compounded by the passage of time. Scenarios become useful in the zone *beyond* the limits of standard prediction. However, we question whether it is really the case that scenarios must only start where prediction ends. Do most organisations that invest in scenario planning have no expectation that the scenarios created will provide a credible picture of what the future might be like? Instead, we view scenarios as anticipatory systems with a distinctive type of predictive model, where prediction is not based on probability, but plausibility. The zone of plausibility is a space in which scenarios provide evidence-based, broad-brush, but targeted, prediction. We contend that other objectives for scenarios, (such as futures consciousness, or re-perception) are necessary but secondary to scenarios' predictive work in the zone of plausibility. Importantly, we argue that if we abandon hope that our scenarios will provide a constrained form of prediction, then we fail to harness the potential that scenarios can offer—when they are plausible, rigorous, and robust.

Keywords Limits of prediction · Complexity, uncertainty and prediction · Scenarios and prediction · Properties of scenarios · Plausible scenarios · Turbulence, Uncertainty, Novelty and Ambiguity (TUNA conditions)

3.1 Overview

> All who claim to foretell or forecast the future are inevitably liars, for the future is not written anywhere—it has still to be built (Godet and Roubelat, 1996, p. 164).

Scenarios are used by a range of disciplines and in a variety of settings for considering futures that lie beyond the limits of standard prediction. Scenario exercises also help us to understand the present. The limits of prediction are defined by three core factors: *complexity*, *uncertainty,* and *the passage of time*. If we grasp the operation of these three dynamics, we will be better placed to choose scenario methods that will help us navigate them skillfully. Here, the aim is to select methods to help us steer a middle course between fatal over-simplification and analysis paralysis.

J. McGee et al., *The Future of Antarctica*, Springer Polar Sciences,
https://doi.org/10.1007/978-981-16-7095-4_3

This chapter investigates the broad purpose and aims of 'scenario planning'. It provides a foundation for later chapters, in which we apply these findings to the specific field of Antarctic geopolitics. We address two related questions: in the context of Antarctic geopolitics, why should we consider scenario planning? and, why is the choice of scenario method important?

3.2 What Are Scenarios?

Our search of the literature on scenario analysis, scenario planning, forecasting and futures has unearthed at least 50 concepts associated with the term 'scenarios'. Many of these have been surveyed and catalogued in various ways (Amer et al. 2013; Chermack 2018; Ramírez et al. 2015; Huss 1988; Bradfield et al. 2005). Before assembling a definition, it is useful to understand what scenarios are *for*, and to identify some of their key properties.

The central idea here is that scenarios become useful as we approach the limits of prediction. If we understand the complementary role that scenarios can play in relation to predictive models, then we can make informed choices about the most appropriate approach (or combination of approaches) for attempting to navigate the future, in any given setting. So, what are the respective roles that scenarios and predictive models might play in an organisation analysing the future? We suggest that scenarios become useful as our analysis approaches the *limits of prediction*. This is the threshold beyond which predictive techniques rapidly lose their capacity to predict accurately. The analyst will have little need for scenarios when all the key variables in a given system can be represented mathematically, and when quantitative models can provide a replicable means to test the 'evolution of a system under alternative assumptions' (O'Mahony 2014, p. 44). Prediction (by extrapolation of existing trends) is more useful in this context.

However, despite increases in the sophistication of modelling and data-intensive analysis, predictive techniques continue to be of limited value for many types of medium- and long-term strategic planning (Huss 1988). Why is this? The main reason is that some conditions facing the contemporary long-range planner lie beyond the current reach of quantified, predictive techniques. What are these conditions? From the literature, it seems the practical limits of predictive models are defined by three core factors: *complexity*, *uncertainty*, and *the passage of time*.

Here, it is important to note that *complexity* means something more than just 'intricate' (Schoemaker 1993) or 'complicated' (Andersson and Törnberg 2018). We cannot grasp how a complex system works just by deconstructing it into progressively simpler sets of components—as we might with a clock. Instead, complex systems are hard (or impossible) to predict. This is because their causal structures overlap (Schoemaker 1993); their components react to each other disproportionately (Byrne and Callaghan 2013); and the overall behaviour of a system cannot be forecast simply by aggregating the behaviour of its constituent parts (Theurer 2014; Kim 1995). In

Chap. 5, we will argue that Antarctic geopolitics are a good example of a complex system.

Navigating complexity in scenario planning is not a trivial undertaking for unskilled hands. In particular, every scenario process must run a formidable gauntlet between: (a) fatal over-simplification, in which key factors and dynamics at work within a system are overlooked; and (b) analysis paralysis, where attempts to process the 'sheer bulk of combinations' (Sharma and Yang 2015, p. 417) bog down the whole exercise, participants disengage, and few useful insights are salvaged. To avoid our scenarios veering into either of these dead ends, the facilitator and core team leading the process need to be given adequate time and the right tools to navigate these various properties of complexity, in the zone beyond standard prediction.

In general usage, *uncertainty* is defined (*inter alia*) as: 'the quality of being uncertain in respect of duration, continuance, occurrence'; 'liability to chance or accident'; 'the quality of being indeterminate as to magnitude or value'; 'something not definitely known or knowable'; and 'a doubtful point' (Oxford Dictionary and [Online] 2021). Uncertainty is distinguished from *risk*:

> *Risk* is the situation under which the decision outcomes and their probabilities of occurrences are known to the decision-maker, and *uncertainty* is the situation under which such information is not available to the decision-maker. (Park and Shapira 2017, emphasis added)

Applying this specifically to scenarios, we adopt Schoemaker's definition of uncertainty as concerning 'the extent to which the causal structure of a strategically relevant variable is *unknown*' (Schoemaker 1993, p. 197). We suggest that at least some of these causes of uncertainty can be reduced if scenario processes are able to access relevant data, select appropriate analytical methods, and engage participants with the requisite knowledge and skills.

The third limit of prediction is *time*. The vitiation of prediction by complexity and uncertainty is compounded further by the passage of time. This is because, except in a static state system, a longer timeframe introduces a greater number of variables. As a result, quantitative, predictive models arguably become less effective and scenarios more effective as the time-period under consideration lengthens (Amer et al. 2013).

Figure 3.1 depicts the impacts of complexity, uncertainty, and the passage of time on the limits of prediction. For illustrative purposes, *complexity* and *uncertainty* are combined as a single variable on the vertical axis.[1] *Time* is represented on the horizontal axis. The dotted red line demarcates the zone beyond the limits of standard prediction, where scenarios may become useful. The basic premise is that we reach the limits to prediction much sooner if our driving forces are complex and uncertain.

So, within this model, how will driving forces behave over time?

- Driving forces that initially have **low** levels of complexity and uncertainty tend to be suited to predictive models well into the long term. Nonetheless, over time they will gradually increase in their complexity and uncertainty (unless they are

---

[1] Loosely adapted from Schoemaker (1993), p. 198, which places complexity and uncertainty on the horizontal and vertical axes.

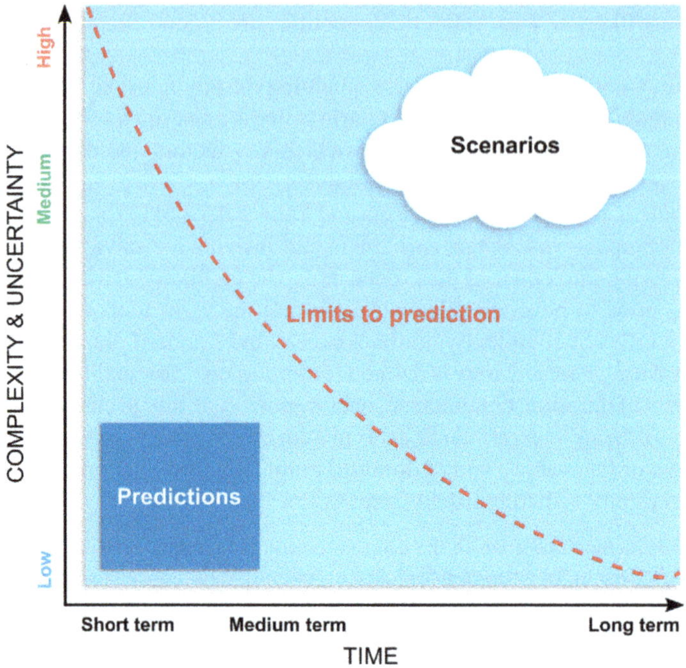

Fig. 3.1 Limits of prediction due to complexity, uncertainty and the passage of time

variables that remain constant). In the long term, there may come a point when predictive models become increasingly inaccurate—we will have reached the limits to prediction. Here, scenarios can play a *supplementary* role in the zone beyond the limits of prediction.

- Driving forces that initially have **medium** levels of complexity and uncertainty will reach the limits of prediction much sooner. This is because there are a larger number of uncertainties and complexities at work, and so the compounding impact of time is more aggressive. In this case, predictive models may play an important role in the short term, but scenarios will become useful from the medium term onwards.
- Finally, driving forces that initially have **high** levels of complexity and uncertainty will pierce the limits of prediction quite quickly. They may be suited to predictive models only in the immediate short term, if at all. In this case, scenarios can play a significant role in organisational planning from the short to medium term onwards.

So far, we have seen how scenarios can be useful in the zone beyond the limits to prediction. Next, we turn to consider their core *properties*; that is, their acknowledged qualities or attributes (Martin 1995). We argue, as do others, that, to be *useful*, scenarios must deliver persuasive inputs for the development of organisational strategy (Ramírez et al. 2015; Ramírez and Wilkinson 2016; Fink et al. 2010;

Lindgren and Bandhold 2009; Nicosia 2011; Richardson 2011). However, we observe that not all the scenario properties identified below will lead to this sort of strategic input. And so, we suggest that some of these properties are best viewed as necessary enablers of rigorous scenario development—but they are not sufficient in themselves. Why is this important? Because if we start out with low expectations of our scenarios, then we miss out on the insights that a rigorous process can deliver.

3.3 Ten Key Properties of Scenarios

1. *Plausible stories about alternative future states*
   Scenarios are *not* forecasts (Chermack 2018), or extrapolations of current trends (Schwartz 1996). Instead, they are small sets of diverse, but internally consistent and equally plausible narrative descriptions (Chermack 2018; Huss 1988) about alternative future states (Schwartz 1996; Nygrén 2019). They aim to highlight the implications should certain trajectories unfold (Godet 1986)—for an organisation, system, or area under focus (Nygrén 2019). Scenarios must be plausible but, taken individually, each is highly unlikely to unfold in the future exactly as described (Derbyshire and Wright 2017).
2. *Causal pathways to future states*
   A scenario is not just a static portrait of possible conditions at some point in the future (Schoemaker 1993). It also involves the construction of a hypothetical sequence of intermediate causal events (the paths, dynamics, and decision points) that could lead us from present conditions to a particular future state (Amer et al. 2013; Huss 1988; Schoemaker 1993; Nygrén 2019; Spaniol and Rowland 2018; Freeth and Drimie 2016; Dannenberg and Fischer 2017). To be plausible, these future paths should not be mere flights of imagination. Instead, their starting points should be sufficiently anchored to factual inputs about present events and trends (Leseure 2019).[2]
3. *'Bounding' the range of future possibilities*
   A scenario cannot prepare us for every possible issue and eventuality. Instead, one property of scenarios is the 'bounding' or delimitation of scenario conditions (Barker 2014), uncertainties (O'Mahony 2014; Bradfield et al. 2016) and narratives to a 'zone of possibilities' (Schoemaker 1993, p. 195), or an 'envelope of probability' (O'Mahony 2014, p. 43). Ideally, these narratives should be fundamentally different from each other, but exhibit plausibility and internal coherence (Schoemaker 1993; Derbyshire and Wright 2017). One organising principle for the bounding of scenarios is to identify internal consistencies and exclude inconsistencies in the causal narrative (O'Mahony 2014). However, bounding based on consistency should not be taken to extremes,

[2] See also, Leonhardt van Efferink, 'Geopolitical Scenario Planning', Course Notes, Maastricht University Summer School 2019, unpublished.

because in the real world, inconsistencies do arise in highly complex environments (O'Mahony 2014). Critically, decisions about scenario bounding are not a value-neutral technical exercise: they will be shaped by the knowledge, skills, experience, values, assumptions, perceptions, and imagination of participants; by the data available; and by the analytic techniques employed. For these reasons, decisions about bounding need to be conscious, explicit, and well-articulated.

4. *Structured thinking about alternative futures*
   Scenarios provide a disciplined ordering of our perceptions (Ramírez et al. 2015). This is achieved through structured analysis of the forces, events, actors, trends, and causal relations (O'Mahony 2014; Meissner et al. 2017)[3] that drive the physical and social systems under review. We will survey some qualitative and quantitative approaches to structured analysis in Chap. 4. For now, the key point is to carefully distinguish this kind of assessment from forecasts (Chermack 2018), which are based on the probabilistic extrapolation of trends, and which, as we have seen, are of limited use in the zone beyond standard prediction.
5. *Capturing complex dynamics*
   It will not be enough if scenarios *only* explore a single dimension of a system or field in isolation (Chermack and Coons 2015).[4] Instead, scenario processes need to capture the key internal and external dynamics, forces, and trends at play within a system or field, and (critically) the *interactions* between those dynamics (Chermack and Coons 2015). Not all these dynamics will be immediately obvious, and failure to model these can leave scenario participants blind to essential complexities with potential to shape future events. It is worth noting that, particularly for social systems, constructing a model of these dynamics is not value-neutral, and will be influenced by the perceptions and assumptions of those involved (Chermack and Coons 2015).
6. *Handling uncertainty*
   A valuable property of scenarios is their capacity to help us navigate environments of uncertainty without resorting to probability-based forecasts (Godet 1986). A vital aspect here is to invest time in distinguishing between key elements in our scenarios that are critical uncertainties, and those that are predetermined. We will unpack this distinction further in Chaps. 4 and 5.
7. *Devices for imagining possible and preferred futures*
   Imagination is a companion property to structured thinking in scenarios. Imagination is 'the power of the mind to consider things which are not present to the senses', often through the recombination of known elements, using mental imagery (Martin 1995, p. 395). Imagination benefits scenarios in three ways. The first benefit is that creative imagination helps to us to 'test the limits of

[3] See also, van Efferink, 'Geopolitical Scenario Planning'.

[4] However, there are methods to capture a general macro-level perspective first, then move to a specific micro-level view (Amer et al. 2013; Fink et al. 2010; Bradfield et al. 2016).

thinking' (Meissner et al. 2017, p. 6), and to cast off the blinkers of incrementalism. A flight or leap (Chermack 2018) of imagination, to 'think the unthinkable'[5] is needed if we are to move beyond a mindset in which the only future paths that are conceivable are probabilistic projections from present conditions (Taleb 2010). An incremental mindset becomes progressively less useful the further we move into the zone beyond the limits of standard prediction. As a second benefit, imagination is useful for the construction of those scenarios that are 'normative' in approach. This is where participants are asked to envision a desired future state that may be quite different to present conditions, and then construct a plausible causal path from there back to the present (Godet and Roubelat 1996; Amer et al. 2013; Ramírez and Wilkinson 2016; Spaniol and Rowland 2018; Godet 1986; Durance and Godet 2010).[6] We will encounter normative scenarios again in Chap. 4. A third benefit of imagination is to excite participant engagement. Here, we draw on imagination to paint the unfolding scenarios in sufficiently vivid detail to trigger visceral connections with the narratives, and so increase the authenticity of participant decision-making under these simulated conditions (Baena 2017).

Wack (writing in 1986) noted the result when imaginative engagement is absent:

> [A] key problem with scenario planning [is that] the interface of scenarios and decision makers is ignored or neglected. By interface, I mean the point at which the scenario really touches a chord in the manager's mind - the moment at which it has real meaning for him or her. The fact that those with the responsibility for preparing the scenario do not feel any responsibility for the interface is the main reason that - despite the logical appeal scenarios should have for managers disenchanted with forecasts - scenario planning has been scarcely developed. (Wack 1985a, p. 139)

Of course, this is a matter of balance, and these exercises of imagination must be sufficiently tethered to empirical data and structured analysis, if our scenarios are to provide more than mere 'entertainment' (Godet and Roubelat 1996).

8. *Reperceiving and challenging current assumptions*
The concept of 're-perceiving' (Ramírez and Wilkinson 2016) is perhaps the second most critical property of scenarios, after the handling of uncertainty. Articulated by Wack (Wack 1985a, b) and elaborated by others (Ramírez and Wilkinson 2016; Lindgren and Bandhold 2009; Schwartz 1996), the central idea is that effective scenarios will include a learning process (Chermack 2018) that challenges participants' framing of issues (Wack 1985a), their conventional wisdoms (Chermack 2018), and their mental models and assumptions about their present reality (Schwartz 1996).[7] If we are prodded to question—and possibly alter—our current mindset, we will not only be more receptive to new phenomena, but will also 'reperceive' familiar phenomena in new ways

---

[5] van Efferink, 'Geopolitical Scenario Planning'.

[6] Using a deductive approach, we define a set of premisses about a plausible future state, then *work backwards* from there to the present. See Chap. 4.

[7] See also, van Efferink, 'Geopolitical Scenario Planning'.

(Wack 1985a). These insights, however unsettling, will help us to make more nuanced and informed decisions about the future (Schwartz 1996). We can do no better than relay some of Wack's insight and experience of the value of reperceiving:

> As Peter Drucker said, 'The greatest danger in times of turbulence is not the turbulence; it is to act with yesterday's logic'. (Drucker, quoted in Wack, 1985a, p. 150)
>
> Highly relevant information goes unnoticed because, being locked into one way of looking, managers fail to see its significance. (Wack 1985a, p. 150)
>
> During stable times, the mental model of a successful decision maker and unfolding reality match. Some adjustment and fine tuning will do. Decision scenarios have little or no leverage. In times of rapid change and increased complexity, however, the manager's mental model becomes a dangerously mixed bag: rich detail and understanding can coexist with dubious assumptions, selective inattention to alternative ways of interpreting evidence, and illusory projections. In these times, the scenario approach has leverage and can make a difference. (Wack 1985a, p. 150)
>
> A mere high or low around a baseline can never achieve a conceptual reframing. The reperception of reality and the discovery of strategic openings that follow the breaking of the manager's assumptions (many of which are so taken for granted that the manager no longer is aware of them) are, after all, the essence of entrepreneurship. (Wack 1985a, p. 150)

9. *Informing strategic planning*
   The construction of scenarios is rarely an end in itself (Leseure 2019). Instead, scenarios are best viewed as intermediate products whose primary function is to provide inputs to strategic planning (Leseure 2019). Specifically, scenarios provide a mechanism for an organisation to stress test the resilience of its current strategies through a hypothetical 'wind tunnel' of alternative future conditions (Chermack 2018; Konno et al. 2014a; Ramírez and Selsky 2016), which may include 'possible, relevant and challenging future changes'. (Ramírez and Selsky 2016, p. 98)

   > …as turbulence threatens to bring forth more unpredictable uncertainty than that which an organization or an industry has had to address and can cope with in non-turbulent conditions, its strategists may want to appreciate what it might be like to conduct business in different versions of the future. (Ramírez and Selsky 2016, p. 98)

   We now turn to the final element in this brief survey of key properties of scenarios.

10. *Enhancing knowledge*
    A significant feature of the scenario process is organisational learning (Ramírez and Wilkinson 2016; Lindgren and Bandhold 2009; Schwartz 1996; Wodak and Neale 2015; Konno et al. 2014b). Indeed, Schwartz has said that 'scenarios are not predictions … [r]ather, scenarios are vehicles for helping people learn' (Schwartz 1996, p. 6). Undoubtedly, one cannot produce useful scenarios that inform strategy unless organisations and individual participants gain a deeper appreciation of their current conditions, as well as the forces and dynamics likely to operate in the future. In addition, a single scenario process is only a

point-in-time snapshot. Ideally, it will be supplemented with further iterations and commitment to an ongoing organisational capability of continuous reorientation of strategy to the future (Ramírez and Wilkinson 2016; Lindgren and Bandhold 2009; Schwartz 1996). However, we suggest that *if that is all* we expect scenarios to do, then we are letting the process 'off the hook' in terms of the methodological rigour that is possible.

In summary, this section has sketched ten properties of scenarios that emerge from the literature. As noted at the beginning, these properties may be thought of as a set of diverse analytical approaches to achieve one or more desired perspectives about possible future states. Taken together, these general properties and specific usages provide a rich picture of the varied roles that scenarios can perform.

3.4 Reality Check: Scenarios and Our Need to Predict

Having surveyed what the literature says about the attributes of scenarios, we question whether it is *really* the case that scenarios only start where prediction ends. Evolving since the 1970s, scenario planning has become a well-used technique within business and government (Lindgren and Bandhold 2009). A major objective cited for an organisation undertaking scenario planning is to 'future-proof' their strategies by anticipating plausible futures, and to improve decision-making (Martelli 2001). To us, it seems unlikely that most organisations would be prepared to divert time and resources to scenario planning *without any expectation* that the scenarios created would provide them with a credible picture of what the future might be like. Drawing on biologist Robert Rosen's anticipatory systems theory, Fuller observes that 'living agents act on prediction: on inferential entailments' (Fuller 2017, p. 47). To survive, organisms (and, by extension, organisations) must run a constant anticipatory paradigm that 'acts faster than the real time of the system' (Fuller 2017, pp. 42–43). However, the sort of 'prediction' deployed in anticipatory systems (such as scenarios) is not the trend-based, extrapolative type that wilts in the face of uncertainty, complexity, and time. Instead, as Louie notes:

> The predictive model in an anticipatory system must not be equivocated to any kind of 'certainty' (even probabilistically) about the future. It is, rather, an assertion based on a model that runs in a faster time scale. The future still has not yet happened: the organism has a model of the future, but not definitive knowledge of future itself. (Louie 2010, p. 20)

We believe it helps to view scenarios as anticipatory systems, with a distinctive type of predictive model, where prediction is not based on probability. What, then, is the basis of a model that provides us with something less than probabilistic prediction, and something more than no future guidance at all? For this, we turn to the concept of *plausibility*, via the famous 'aha' moment described by Pierre Wack.

3.5 Plausibility and the 'aha' Moment

Here, we suggest that the holy grail of scenario planning—the much sought-after 'aha' moment (Schwartz 1996; Wack 1985a; van der Heijden 1996) of reperception (Ramírez and Wilkinson 2016; Wack 1985a)—requires that our scenarios must be *plausible*. In describing this 'aha' moment, Wack observes that effective scenarios first break, then reset, the mindset of decision makers:

> I have found that getting to that management 'Aha!' is the real challenge of scenario analysis. It does not simply leap at you when you've presented all the possible alternatives, no matter how eloquent your expression or how beautifully drawn your charts. It happens when your message reaches the microcosms of decision makers, obliges them to question their assumptions about how their business world works, and leads them to change and reorganize their inner models of reality. (Wack 1985a, p. 140)

An 'aha' moment is a new, actionable realisation that builds forward, with credibility, from existing understanding. As Schwartz notes, the 'aha'—this change of inner models—may lead organisations to alter their plans drastically (Schwartz 1996). However, for leaders to have sufficient confidence to change strategy, redirect resources and embrace new risks, it is crucial that the scenarios created be plausible. Indeed, plausibility is considered one of the core benchmarks for scenario success (Amer et al. 2013; Chermack 2018; Bradfield et al. 2005; Lindgren and Bandhold 2009; Schwartz 1996; Bezold 2009; Bishop 2017). Within the scenario literature, however, the term is often used without being defined.

So, what exactly do we mean by 'plausibility'? To start with, Lindgren and Bandhold draw a useful distinction between plausibility and probability. 'Probability' is generally assessed over the short term and is concerned with the *likelihood* that future events will happen, that a scenario will be realised (Lindgren and Bandhold 2009). In contrast, 'plausibility' is generally judged over the longer term and is concerned with whether proposed events are both theoretically *possible* and *feasible.*

In the context of scenarios, plausibility has a number of attributes. To be plausible, scenarios must be realistically possible (Lindgren and Bandhold 2009); capable of happening (Amer et al. 2013); what might happen (Bezold 2009); credible (Amer et al. 2013); 'more than just theoretically possible' (Bishop 2017, p. 87); and 'more probable than mere possibility for which there is no reason or evidence' (Bishop 2017, p. 88). There should also be 'reasons (evidence) to suggest that they might happen' (Bishop 2017, p. 88). Ramírez and Wilkinson note that plausibility 'seeks to relate novelty in a productive tension with assumptions derived from historical determinism' (Ramírez and Wilkinson 2014, p. 256). In other words, by applying a test of plausibility to our scenarios, we form a bridge of credibility between the empirical evidence we have about the present and our imagination about possible futures. Plausible scenarios should 'bound the range of uncertainties that appear inherent in the future' (Bradfield et al. 2016, p. 60); they should also be comprehensive, taking into account all relevant variables and trends (Gambelli et al. 2010), including assessing the plausibility of combinations of key factors and critical uncertainties (Amer et al.

2013). Finally, it has been suggested that plausible scenarios also have attributes of clarity, coherence, and consistency (Gambelli et al. 2010).

So, how do we know if our scenarios have achieved plausibility? If scenarios are to challenge prevailing assumptions and craft compelling narratives, then participants need the space (and licence) to imagine, dream, and think 'outside the box' (Chermack 2018; Ramírez and Wilkinson 2016; Schwartz 1996). At the same time, if these narratives are to be plausible, then mere dreaming is not enough—these stories about alternative potential futures need grounding in evidence. Here, we agree with Bishop and challenge the post-modernist view that plausibility of scenarios is self-evident (Bishop 2017). We suggest that plausibility needs to be 'evidence-based'—by which we mean grounded in the rigorous (and judicious) use of empirical data and sound reasoning, with clarity about assumptions, and capable of withstanding scrutiny and debate (Bishop 2017). It is hard to imagine that decision makers will be prepared to change course, at least on matters of significance, without such an evidence base.

Before going further, we note that plausible scenarios tread a narrow path between two hazards. The first is the *hazard of invention,* where we conflate: (a) logical reasoning, with (b) empirical evidence. This where we construct a hypothetical, but logical, sequence of future events, based primarily on flights of imagination, which is *tethered too loosely* to an evidence base about current conditions. The stories created may have a beguiling rationality, as one step leads to the next, but they fail a key test for plausibility: there is no reason to suggest the proposed outcome is anything more than theoretically possible. Organisational leaders are unlikely to experience a compelling 'aha' moment if scenarios are not sufficiently connected to their existing understanding. In other words, there must be a credible line of sight between the future drama that unfolds within a scenario and the gritty realities of present 'facts on the ground'.

The second, opposite, danger is the *hazard of extrapolation*, where we conflate: (a) a solid evidence base about the present, with (b) our ability to predict the future. This is where we construct a hypothetical sequence of future events, based primarily on extrapolating from the present into the future, which is *tethered too tightly* to an evidence base about current conditions, and which is unchallenged by imagination. Again, these stories may have a beguiling rationality as we work through various forms of data and analysis but, as noted above, they risk lulling us into a false sense of probabilistic certainty (Louie 2010). So, in summary, the evidence base for plausible scenarios needs to provide us with something more than logical coherence, and something less than probability.

What, then, is the evidence base for plausible scenarios? We unpack this further in Chap. 4. For now, it is enough to be aware of the breadth of approaches to data and analysis, when building an evidence base. These include, for example, critical interpretation (Ramírez et al. 2015); historical events with ongoing influence[8]; judgement and statistical inferences from past events (Nicosia 2011; Mack 2010; Suharto 2013); current baseline events and critical variables (O'Mahony 2014; Konno et al. 2014a); weak signals of a potentially substantial future change (Gambelli et al. 2010);

[8] While acknowledging the contingent nature of causation in historical events (Suharto 2013).

consensus of expert opinion (Suharto 2013; Kaufmann 2016; Agustan et al. 2019; Hernandez et al. 2019; United Nations Industrial Development Organization 2005; Ritchey 2011); ruling out mutually exclusive/implausible combinations of factors (Nygrén 2019; Lord et al. 2016) and integration of multiple scenarios (Barker 2014). The strength of scenario plausibility depends, in part, on the rigour of the evidence base. We suggest that rigour is more likely to be achieved if we employ a judicious balance of the approaches listed above.

At this point, it may help to restate the argument so far. We began by suggesting that scenarios are useful in the zone of the future that lies beyond the limits of standard prediction, and that the limits of prediction are defined by three core factors: *complexity*, *uncertainty*, and *the passage of time*. Then, having surveyed what the literature says about the attributes of scenarios, we questioned whether it is *really* the case that scenarios only start where prediction ends. We suggested that the holy grail of scenario planning, the much sought-after 'aha' moment, requires that our scenarios be *plausible*. To be plausible, scenarios need to be both possible and feasible. They must also provide us with something more than logical coherence, but something less than probability. This demands that scenarios be grounded in, but not unduly constrained by, an evidence base. Ideally, this evidence base will draw on a variety of approaches to data and analysis. With these insights, we now need to revise our initial model. In particular, we need to indicate that scenarios require plausibility in the zone beyond standard prediction.

3.6 A Revised Model: Introducing the Zone of Plausibility

Here, we coin the term *zone of plausibility*. This is shorthand for the space in which scenarios provide *evidence-based*, *broad-brush*, but *targeted* prediction, of the type sought by organisations in the zone beyond the limits of standard prediction. We have defined 'evidence-based' in the preceding sections. With 'broad-brush', we adopt standard usage: '(as if) painted with a broad brush; wide-ranging but lacking in detail; general, sweeping' (Oxford Dictionary [Online] 2021). Here, we follow van der Heijden (van der Heijden 1996, p. 95) and Derbyshire and Wright (Derbyshire and Wright 2017, p. 254) in using this phrase to depict the kind of knowledge and decision-making that is possible in the zone of plausibility. Painting with a broad brush avoids the *hazard of extrapolation* because we do not bind ourselves too tightly to an incremental progression from current conditions. To avoid the opposite *hazard of invention*, however, our scenarios should not admit any outcome just because it is theoretically possible. So, here, broad-brush prediction is tempered by being targeted. By 'targeted', we refer both to: (a) the intentional shaping of scenarios to encompass topics of strategic interest to the target audience; and (b) the bounding of scenarios within those realms of future possibilities that are plausible (Barker 2014). In summary, we suggest that these three criteria (evidence-based, broad-brush, targeted) reduce the theatre of operations for scenarios to a more limited region within the zone beyond standard prediction, which we call the *zone of plausibility*.

Where does this leave other scenario goals, such as ‘reperceiving’ (Ramírez and Wilkinson 2016; Lindgren and Bandhold 2009; Schwartz 1996; Wack 1985a, 1985b)? We contend that other, often-stated objectives for scenarios (such as improving an organisation’s futures consciousness (Ahvenharju et al. 2018), or changing its mindset (Chermack 2018, p. 52)) are necessary, but secondary to scenarios’ predictive work in the zone of plausibility. In many cases, they are essential means to that end but, if we abandon any expectation that scenarios can provide a constrained type of predictive accuracy, then we unnecessarily reduce our expectations for methodological rigour in scenario planning. In so doing, we fail to harness the potential that scenarios offer, and are left with little more than ‘entertainment’ (Godet and Roubelat 1996).

So, how do we recognise when our scenarios have delivered this constrained form of prediction? Here, we suggest a two-fold test:

- The first criterion is that organisations undertaking a scenario process expect that the realised future will fall somewhere within the broad directions depicted in the basket of scenarios (taken as a whole). However, we note that this can only be validated retrospectively.
- The second criterion is that the basket of scenarios should not be so broad and so general that it provides no direction for the evaluation of current strategy.

These criteria are depicted in Fig. 3.2. This illustrates the relationships between the zones of the *predictable*, the *plausible,* and the *possible.* If our scenarios are to provide evidence-based, broad-brush, but targeted prediction, then they will fall outside the zone of standard prediction (dark blue), but within the zone of plausibility (mid-blue).

To meet the first criterion above, the realised future will fall somewhere within the total basket of scenarios produced, *when taken as a whole.* This is the sum of eventualities covered by multiple scenarios, regardless of whether they are ultimately integrated (see Chap. 4) or left to stand independently of each other. This red square, then, delimits the zone within which an organisation’s future strategy is directed. To meet the second criterion, the basket of plausible scenarios will have (ideally) a much narrower focus than the sum of all possible futures, which is represented by the larger, light blue square. This targeted focus avoids the creation of scenarios that are too general to be of any practical guidance.

Taken together, these two criteria may help us to recognise when our scenarios are delivering what organisations expect. However, we must avoid creating the ‘illusion of certainty’ (O’Mahony 2014, p. 42) and ensure that both developers and users of the scenarios understand the constrained meaning conveyed by this type of ‘prediction’. We are mindful that a call for scenarios to provide any form of predictive function runs counter to a good portion of the scenario literature. As one example, Ramírez and Wilkinson take the position that:

> We think that prediction is *not* a good criterion to determine if scenario planning worked well; if prediction were possible, scenario planning has no relevance, and is not needed. [Ramírez and Wilkinson 2016, p. 114, emphasis in original]

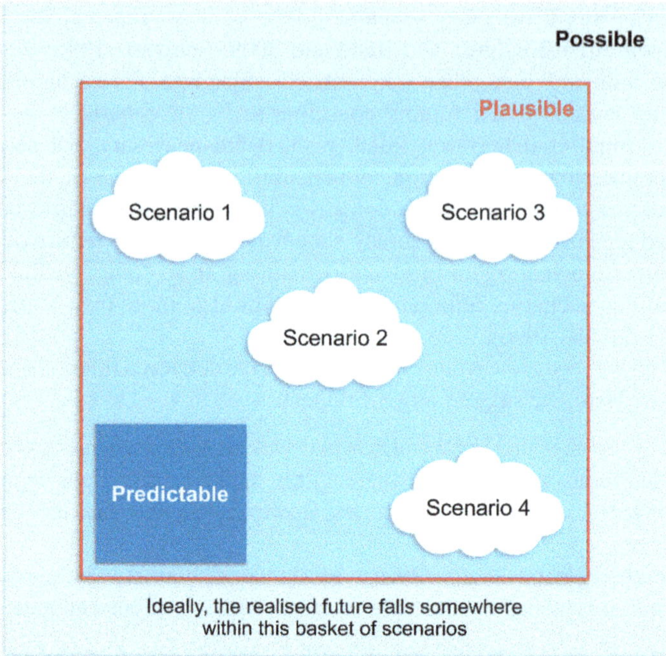

Fig. 3.2 Zones of the predictable, the plausible, and the possible

In response, we agree that the value of extrapolative prediction diminishes as we approach the limits of standard prediction. However, we suggest that evidence-based, broad-brush, but targeted prediction is somewhat different to conventional notions of extrapolative prediction. Further theoretical work and empirical validation is needed before we would claim to be on firm ground.

3.7 Dissecting Turbulence, Uncertainty, Novelty and Ambiguity (TUNA)

How does our depiction compare with other models in the literature? Here, we have space to engage with just one—the influential Oxford Scenario Planning Approach (OSPA) (Ramírez and Wilkinson 2016). There, Ramírez and Wilkinson suggest that scenarios become useful when organisations encounter increasingly prevalent conditions of Turbulence, Uncertainty, Novelty and Ambiguity (TUNA conditions).

- *Turbulence* is perhaps the least familiar and draws on Emery and Trist's work on causal textures theory in organisational environments (Ramírez and Wilkinson 2016). It refers to an environment where a high level of complexity is aggravated by a high rate of change. This leads to increased uncertainty and produces an 'unsettling' 'texture of turbulence' (Ramírez and Wilkinson 2016, p. 8). The risk for strategic planning is that 'conditions in the *contextual environment* overcome the *transactional environment* and prevailing ground rules among actors...no longer apply or are questioned deeply'; and all this compromises an actor's perceived capacity to adapt (Ramírez and Wilkinson 2016, p. 221, emphasis in original).[9]

The remaining TUNA conditions are relatively straightforward.

- *Uncertainty* is contrasted with calculable risks, and refers to unpredictable events and states which cannot be measured, but can be imagined (Ramírez and Wilkinson 2016).
- *Novelty* refers to unprecedented situations in human history, often requiring new vocabulary and units of analysis (Ramírez and Wilkinson 2016 following Emery and Trist).
- *Ambiguity* occurs where an event or phenomenon can give rise to different, potentially irreconcilable, interpretations or framings (Lindgren and Bandhold 2009).

While we agree that each of the TUNA concepts is important, we question whether they reflect the most fundamental categories of the strategic planning landscape. In particular, we question the relegation of *complexity* to a sub-category of *turbulence* under the TUNA model (Ramírez and Wilkinson 2016), and argue that the reverse may be true. For example, it is conceivable that some of the more structural attributes of complexity—such as 'massive parallelism' andmulti-level hierarchies (Andersson and Törnberg 2018, p.119)—can occur quite independently of turbulence. If so, then complexity cannot be a sub-category of turbulence. Why does this matter? We suggest this is important because, if our scenario planning focuses primarily on the narrower condition of turbulence, rather than on the broader notion of complexity, then we may overlook other attributes of complexity that may have pivotal roles as our scenarios unfold.

3.8 Summary

In this chapter, we have suggested that the limits of prediction are defined by three core factors: *complexity*, *uncertainty,* and *the passage of time*. We began with the premise that scenarios become useful as we approach the limits of prediction. We noted ten key properties of scenarios, as identified in the literature, which led us to question whether it is *really* the case that scenarios only start where prediction ends. We feel it is unlikely that most organisations would be prepared to divert time and resources to scenario planning without any expectation that the scenarios created would provide them with a credible picture of what the future might be like. We believe it helps to

[9] See also Young's related concept of endogenous/exogenous misalignment within institutions (Young 2010).

view scenarios as anticipatory systems with a distinctive type of predictive model, where prediction is not based on probability, but plausibility. We suggest that to be useful, scenarios need to operate within the *zone of plausibility*. This is the space in which scenarios provide evidence-based, broad-brush, but targeted, prediction. We contend that other, oft-stated objectives for scenarios, (such as futures consciousness, or changing mindsets) are necessary but secondary to scenarios' predictive work in the zone of plausibility. Further, we argue that if we abandon hope that our scenarios will provide a constrained form of prediction, then we unnecessarily reduce our expectations for methodological rigour. In so doing, we fail to harness the potential that rigorous scenario processes can offer. We explore this potential in Chap. 4.

It is worth re-emphasising that a single scenario process is only a point-in-time snapshot. Ideally, it will be supplemented with further iterations and a commitment to continuous reorientation of organisational strategy to the future.

A final observation. The literature on scenarios is rich, diverse and 'furiously conflicting' (Martelli 2001, para 19). It is also 'crowded with contradictions and paradoxes' (Lindgren and Bandhold 2009, p. 44). We suggest that scenario planning suffers from a form of 'town and gown syndrome' (Wescoe 1963), in which its practice struggles for scholarly recognition, and its theory eludes practical relevance. In part, this is because the practice of scenario planning has often developed in advance of theory (Amer et al. 2013; Chermack 2018; Wright et al. 2013). This constructivist approach has contributed to Martelli's now-famous critique that scenario planning is in a state of 'methodological chaos' (Martelli 2001, para 19). He notes:

> Scenarios suffer from a lack of paradigms in Kuhn's sense … i.e. they lack a set of theories, principles and practical rules commonly accepted by at least the vast majority of the theoreticians and of the practitioners. (Martleli 2001, para 19)

Martelli's charge has become such a lodestone of scenario planning literature that Spaniol and Rowland conclude we are overrun with the idea (Spaniol and Rowland 2018). On this view, scenario theory is dismal because it is not 'tethered to a common core' and lacks a 'shared foundational perspective' (Spaniol and Rowland 2018, pp. 33–34). For example, some scenario practitioners adopt the classical scientific method, while others do not (Martelli 2001). The methodological disarray incites a proliferation of books (like this one) to organise the field. These efforts inject 'theory after theory' and 'typology after typology' (Spaniol and Rowland 2018, p. 34), so that, paradoxically, the goal of unified scenario theory moves further away (Spaniol and Rowland 2018). However, while Martelli's charge of methodological chaos is undoubtedly correct, we wonder whether it is reasonable to expect methodological *order*?

We suggest that seeking a unified theory for scenarios offends the principle of theoretical parsimony. Here, we wield Ockham's Razor to assert that seeking a unified/systematic theory (Grassl 2017; Umbrello 2018) or holistic model for scenario planning generates unnecessary complexity. It offends the principles of 'parsimony' in the construction of theory (Pacer and Lombrozo 2017) and minimalism in ontological commitments (Long 2019). We suggest that straining for a general theory of scenario planning is possibly unattainable (Wright et al. 2013) and probably unnecessary. In any event, as Martelli notes, practitioners have tended to just get on with it:

> The lack of generally accepted paradigms, though lamented by a few, was not considered as a serious hindrance for a further development of the scenario approach. [Martelli 2001, para 26]

References

Agustan, Djoko Karmiadji, Mega N. Putri, Oni B. Bintoro, Ira N. Djarot, Hasan Soleh, Aris Irawan et al. 2019. Combining Delphi Study and Scenario Planning for Indonesia Research Priorities in Maritime Sector. IEEE International Conference on Aerospace Electronics and Remote Sensing Technology (ICARES). Yogyakarta, Indonesia, 17–18 October 2019, 1–7. New York: IEEE. https://ieeexplore.ieee.org/document/8914342. Accessed 19 July 2021.

Ahvenharju, Sanna, Matti Minkkinen, and Fanny Lalot. 2018. The Five Dimensions of Futures Consciousness. *Futures* 104: 1–13.

Amer, Muhammed, Tugrul U. Daim, and Antonie Jetter. 2013. A Review of Scenario Planning. *Futures* 46: 23–40.

Andersson, Claes, and Petter Törnberg. 2018. Wickedness and the Anatomy of Complexity. *Futures* 95: 118–138.

Baena, Alethea B. M. 2017. Gaming Futures: To Experience Scenarios Through *Teatro Del Devenir*. *Journal of Futures Studies* 22 (2): 119–124.

Barker, Alec D. 2014. Applying Structured Scenario Fusion to Address Multifaceted Strategic Challenges. *Journal of the Operational Research Society* 65 (11): 1726–1734.

Bezold, Clement. 2009. Aspirational Futures. *Journal of Futures Studies* 13 (4): 81–90.

Bishop, Peter C. 2017. Baseline Analysis: The Epistemology of Scenario Support. *World Futures Review* 9 (2): 83–92.

Bradfield, Ron, George Wright, George Burt, George Cairns, and Kees Van der Heijden. 2005. The Origins and Evolution of Scenario Techniques in Long Range Business Planning. *Futures* 37 (8): 795–812.

Bradfield, Ronald, James Derbyshire, and George Wright. 2016. The Critical Role of History in Scenario Thinking: Augmenting Causal Analysis within the Intuitive Logics Scenario Development Methodology. *Futures* 77: 56–66.

Byrne, David, and Gill Callaghan. 2013. *Complexity Theory and the Social Sciences: The State of the Art*. Abingdon, UK: Routledge.

Chermack, Thomas J. 2018. An Analysis and Categorization of Scenario Planning Scholarship from 1995–2016. *Journal of Futures Studies* 22 (4): 45–59.

Chermack, Thomas J., and Laura Coons. 2015. Scenario Planning: Pierre Wack's Hidden Messages. *Futures* 73: 187–193.

Dannenberg, Sascha, and Nele Fischer. 2017. Gaming Scenarios: Making Sense of Diverging Developments. *Journal of Futures Studies* 22 (2): 15–26.

Derbyshire, James, and George Wright. 2017. Augmenting the Intuitive Logics Scenario Planning Method for a More Comprehensive Analysis of Causation. *International Journal of Forecasting* 33 (1): 254–266.

Durance, Phillipe, and Michel Godet. 2010. Scenario Building: Uses and Abuses. *Technological Forecasting and Social Change* 77 (9): 1488–1492.

Fink, Alexander, Andreas Siebe, and Jens-Peter. Kuhle. 2010. How Scenarios Interconnect Strategy, Innovation, and Early Warning Processes. *World Futures Review* 2 (1): 5–30.

Freeth, Rebecca, and Scott Drimie. 2016. Participatory Scenario Planning: From Scenario 'Stakeholders' to Scenario 'Owners'. *Environment: Science and Policy for Sustainable Development* 58(4): 32–43.

Fuller, Ted. 2017. Anxious Relationships: The Unmarked Futures for Post-Normal Scenarios in Anticipatory Systems. *Technological Forecasting and Social Change* 124: 41–50.

Gambelli, Danilo, Daniela Vairo, and Rafaelle Zanoli. 2010. Exploiting Qualitative Information for Decision Support in Scenario Analysis. *Journal of Decision Systems* 19 (4): 407–422.

Godet, Michel. 1986. Introduction to *La Prospective*: Seven Key Ideas and One Scenario Method. *Futures* 18 (2): 134–157.

Godet, Michel, and Fabrice Roubelat. 1996. Creating the Future: The Use and Misuse of Scenarios. *Long Range Planning* 29 (2): 164–171.

Grassl, Wolfgang. 2017. Toward a Unified Theory of Value: From Austrian Economics to Austrian Philosophy. *Axiomathes* 27: 531–559.

Hernandez, N. Batista, M. Bernarda Ruilova Cueva, B. Narcisa Mazacón, and K. de Mora L. 2019. Prospective Analysis of Public Management Scenarios Modeled by the Fuzzy Delphi Method. *Neutrosophic Sets and Systems* 26(1): 114–119.

Huss, William R. 1988. A Move Toward Scenario Analysis. *International Journal of Forecasting* 4 (3): 377–388.

Kaufmann, Petra Radeljak. 2016. Integrating Factor Analysis and the Delphi Method in Scenario Development: A Case Study of Dalmatia, Croatia. *Applied Geography* 71: 56–68.

Kim, Jaegwon. 1995. Emergent Properties. In *The Oxford Companion to Philosophy*, ed. Ted Honderich, 224. Oxford: Oxford University Press.

Konno, Noboru, Ikujiro Nonaka, and Jay Ogilvy. 2014a. Scenario Planning: The Basics. *World Futures: The Journal of General Evolution* 70 (1): 28–43.

Konno, Noboru, Ikujiro Nonaka, and Jay Ogilvy. 2014b. The Mind of the Scenario Thinker. *World Futures: The Journal of General Evolution* 70 (1): 44–51.

Leseure, Michel. 2019. Is the Recovery of Failing Scenario Analysis a Legitimate and Valuable Activity? *World Futures Review* 11 (1): 80–97.

Lindgren, Mats, and Hans Bandhold. 2009. *Scenario Planning: The Link Between Future and Strategy*, 2nd ed. Basingstoke, UK: Palgrave Macmillan.

Long, Bruce. 2019. A Scientific Metaphysics and Ockham's Razor. *Axiomathes* 29: 483–513.

Lord, Steven, Ariella Helfgott, and Joost M. Vervoort. 2016. Choosing Diverse Sets of Plausible Scenarios in Multidimensional Exploratory Futures Techniques. *Futures* 77: 11–27.

Louie, A.H. 2010. Robert Rosen's Anticipatory Systems. *Foresight* 12 (3): 18–29.

Mack, Ted C. 2010. Memes in Foresight: Response to the Black Swan. *World Futures Review* 2 (2): 5–15.

Martelli, Antonio. 2001. Scenario Building and Scenario Planning: State of the Art and Prospects of Evolution. *Futures Research Quarterly* 17 (2): 57–70.

Martin, Michael G. F. 1995. Imagination. In *The Oxford Companion to Philosophy*, ed. Ted Honderich, 395. Oxford: Oxford University Press.

Meissner, Philip, Christian Brands, and Torsten Wulf. 2017. Quantifying Blind Spots and Weak Signals in Executive Judgment: A Structured Integration of Expert Judgment into the Scenario Development Process. *International Journal of Forecasting* 33 (1): 244–253.

Nicosia, Manuela. 2011. Scenario-Based Strategy: An Interview with Tim Mack, President of the World Future Society. *World Futures Review* 3 (4): 63–70.

Nygrén, Nina A. 2019. Scenario Workshops as a Tool for Participatory Planning in a Case of Lake Management. *Futures* 107: 29–44.

O'Mahony, Tadhg. 2014. Integrated Scenarios for Energy: A Methodology for the Short Term. *Futures* 55: 41–57.

Oxford English Dictionary [Online]. 2021. Oxford: Oxford University Press. https://www.oed.com. Accessed 15 July 2021.

Pacer, M., and Tania Lombrozo. 2017. Ockham's Razor Cuts to the Root. *Journal of Experimental Psychology: General* 146 (12): 1761–1780.

Park, Francis K., and Zur Shapira. 2017. Risk and Uncertainty. In *The Palgrave Encyclopedia of Strategic Management*, eds. Mie Augier and David J. Teece. London: Palgrave Macmillan.

Ramírez, Rafael, and John W. Selsky. 2016. Strategic Planning in Turbulent Environments: A Social Ecology Approach to Scenarios. *Long Range Planning* 49 (1): 90–102.

Ramírez, Rafael, and Angela Wilkinson. 2014. Rethinking the 2×2 Scenario Method: Grid or Frames? *Technological Forecasting and Social Change* 86: 254–264.

Ramírez, Rafael, and Angela Wilkinson. 2016. *Strategic Reframing: The Oxford Scenario Planning Approach*. Oxford: Oxford University Press.

Ramírez, Rafael, Malobi Mukherjee, Simona Vezzoli, and Arnoldo Matus Kramer. 2015. Scenarios as a Scholarly Methodology to Produce 'Interesting Research.' *Futures* 71: 70–87.

Richardson, Jacques. 2011. Wishful Thinking: A Serious Risk Factor in Planning and Strategy? *World Futures Review* 3 (4): 48–62.

Ritchey, Tom. 2011. Modeling Alternative Futures with General Morphological Analysis. *World Futures Review* 3 (1): 83–94.

Schoemaker, Paul J. H. 1993. Multiple Scenario Development: Its Conceptual and Behavioral Foundation. *Strategic Management Journal* 14 (3): 193–213.

Schwartz, Peter. 1996. *The Art of the Long View: Planning for the Future in an Uncertain World*. Chichester, UK: John Wiley & Sons.

Sharma, Ravi S., and Yi Yang. 2015. A Hybrid Scenario Planning Methodology for Interactive Digital Media. *Long Range Planning* 48(6): 412–429.

Spaniol, Matthew J., and Nicholas J. Rowland. 2018. The Scenario Planning Paradox. *Futures* 95: 33–43.

Suharto, Yulianto. 2013. Study of Multi-Scenario Based Technology Roadmapping: Bayesian Causal Maps Approach. *Proceedings of PICMET '13: Technology Management in the IT-Driven Services* (PICMET). San José, California, 28 July-1 August 2013, 2212–2218. Portland, Oregon: PICMET. https://ieeexplore.ieee.org/document/6641847. Accessed 19 July 2021.

Taleb, Nassim N. 2010. *The Black Swan: The Impact of the Highly Improbable*. New York: Random House.

Theurer, Kari L. 2014. Complexity-based Theories of Emergence: Criticisms and Constraints. *International Studies in the Philosophy of Science* 28 (3): 277–301.

Umbrello, Steven. 2018. A Theory of Everything? *Cultural Studies Review* 24 (2): 184–186.

United Nations Industrial Development Organization. 2005. *UNIDO Technology Foresight Manual, Vol. 2, Technology Foresight in Action*. Vienna: UNIDO.

van der Heijden, Kees. 1996. *Scenarios: The Art of Strategic Conversation*. Chichester, UK: Wiley.

Wack, Pierre. 1985a. Scenarios: Shooting the Rapids. *Harvard Business Review* (November-December 1985): 139–150.

Wack, Pierre. 1985b. *Scenarios: The Gentle Art of Re-perceiving: One Thing or Two Learned While Developing Planning Scenarios for Royal Dutch/Shell*. The Hague: Shell International Petroleum Company and Boston, US: Division of Research, Harvard Business School.

Wescoe, W. Clarke. 1963. The Town-Gown Syndrome: Pathology. *Journal of the American Medical Association (JAMA)* 186 (8): 785–786.

Wodak, Josh, and Timothy Neale. 2015. A Critical Review of the Application of Environmental Scenario Exercises. *Futures* 73: 176–186.

Wright, George, and Paul Goodwin. 2009. Decision Making and Planning under Low Levels of Predictability: Enhancing the Scenario Method. *International Journal of Forecasting* 25 (4): 813–825.

Wright, George, Ron Bradfield, and George Cairns. 2013. Does the Intuitive Logics Method—and its Recent Enhancements—Produce 'Effective' Scenarios? *Technological Forecasting and Social Change* 80 (4): 631–642.

Young, Oran R. 2010. *Institutional Dynamics: Emergent Patterns in International Environmental Governance*. Cambridge, MA: MIT Press.

Chapter 4
Objectives, Approaches and Techniques

Abstract From our review of scenario objectives, approaches, and techniques, we suggest that scenario methods can be grouped under one or more of seven broad approaches: deductive, inductive, quantitative, hybrid, integrated, scoping, and engagement. Each has its strengths and weaknesses. We conclude that no single scenario approach can match the scale and complexity of Antarctic geopolitics. For that reason, we recommend that a scenario process employ a judicious blend of several approaches. For this book, we focus on one technique, the standard intuitive logics '2 × 2' matrix, commonly deployed in geopolitical scenarios. We unpack it with a worked example and encounter some difficulties applying it to the field of Antarctic geopolitics. This leads us to propose some modifications, which may be useful for scenarios in other fields. Importantly, 'practical' decisions about scoping and framing of scenarios are shaped, in important ways, by the assumptions and values we hold. Which issues are in? Which are out? How are issues framed? Whose voices are heard, and whose are excluded? What types of data do we admit? How do we handle complexity? What expertise do we need? Decisions about scenario process are not the objective, technical considerations they are often perceived to be.

Keywords Scenario objectives · Scenario approaches · Scenario techniques · Oxford Scenario Planning Approach (OSPA) · Scenario toolbox · Intuitive logics · Scenario matrix

4.1 Introduction

> The gist of the scenario method seems that it is many things: art and science, deduction and induction, structured and fluid, rational (in the unitary actor sense) and political. These multiple facets have caused it to remain elusive and fuzzy by academic standards. Nonetheless, the use of scenarios in strategic management is real, important and growing. (Schoemaker 1993, p. 194)

In the preceding chapter, we described a space where standard, probabilistic predictions are doomed to fail due to the convergence of complexity, uncertainty, and time. Nonetheless, we proposed that, within that space, there is opportunity for scenarios to provide evidence-based, broad-brush, but targeted prediction. We

J. McGee et al., *The Future of Antarctica*, Springer Polar Sciences,
https://doi.org/10.1007/978-981-16-7095-4_4

suggested that that this special type of prediction requires that our scenarios be plausible, and we saw that plausibility has various attributes. We argued that the strength of scenario plausibility depends, in part, on the rigour of the evidence base. And we contended that we are more likely to achieve scenario rigour if we employ a judicious balance of approaches to data collection and analysis. This chapter builds on this work by surveying some of the key approaches available for carrying out a scenario planning process.

As a starting point, we adopt the well-known Oxford Scenario Planning Approach (OSPA) (Ramírez and Wilkinson 2016) to review the deliberate choices we can make in designing scenario planning processes. Then, we develop the concept of a 'scenario toolbox' and suggest that methods can be grouped loosely under one or more of eight broad approaches. These approaches reflect differences in scenario objectives, the types of data available for use in analysis and the forms of reasoning that might be employed. Having set out these key design choices for scenario planning, we focus on a prominent set of techniques grouped under the umbrella of the Intuitive Logics (IL) school. This school was chosen because it is suited to handling the types of qualitative inputs that are common in fields such as geopolitics[1] and because its usage is extensive and well documented.[2] Then, within this school, we narrow our gaze to a single technique that is frequently deployed—the 'IL 2 × 2' matrix (Cairns et al. 2013). This technique was selected primarily because of its clear approach to identifying and framing key issues, and for its potential to tease out differences in scenario participants' understanding, interpretation and agendas (Cairns et al. 2013).

For us, this last point is important—a recurring theme of this book is the need for Antarctic geopolitical scenarios to employ rigour in capturing diverse perspectives. We shall return to this point in Chap. 5. To test some of the strengths and weaknesses of the IL 2 × 2 technique, we unpack the process over eight steps, using a topical example from Antarctic geopolitics—resource extraction. In doing so, we encounter some difficulties with this technique (at least in its standard formulation) which may be relevant for scenario practice in general. This leads us to propose some modifications to the scenario planning process, to provide a more nuanced analysis of driving forces. The IL 2 × 2 technique works very neatly to produce well-contrasted outcomes (within its logic) but if the scenarios are to provide targeted, broad-brush prediction, then a middle option—an additional scenario (Scenario 5)—that describes a medium between both critical uncertainties may be needed. The chapter concludes with an outline of this approach, which we have labelled 'IL 5'.

---

[1] Leonhardt van Efferink 'Geopolitical Scenario Planning', Course Notes, Maastricht University Summer School 2019, unpublished.

[2] For examples, see (Ramírez and Wilkinson 2016; Wright et al. 2013; Huss 1988; Amer et al. 2013; Ramírez et al. 2015; Derbyshire and Wright 2017; Ramírez and Wilkinson 2014; Cairns et al. 2013; Bradfield et al. 2005; Schwartz 1996).

4.2 Objectives, Approaches, and Techniques

Before proceeding, it may help to clarify some key terms. Aspects of scenario practice are described variously in the literature as: objectives, goals, approaches, methods, frameworks, logics, stages, phases, processes, practices, systems, techniques, and tools.

Within the OSPA, Ramírez and Wilkinson label their three descending levels of decision-making as: 'intervention design'; 'choice of method'; and 'tools and techniques' (Ramírez and Wilkinson 2016, pp. 120–121). However, we have found that the terms *method* and *methodology* are used variably across the scenario literature to convey a diverse set of meanings, ranging from high-level philosophical approaches through to detailed techniques. As we shall see, the decisions we make about general approaches are frequently not of the same kind as the decisions we make about specific techniques—and, ideally, we will make informed decisions about both. For this reason, and to avoid the ambiguity exhibited in the term 'method', for the purpose of this book we standardise the various usages noted above into a three-layered terminology that parallels (but only roughly) the OSPA stages: strategic *objectives*; general *approaches*; and specific *techniques*. In doing so, we acknowledge that these are not discrete categories, and that a 'method' described in the literature will often include aspects of both general approach and specific technique.

We use *objective* in its standard sense of being '[a] thing aimed at or sought; a target, goal, or end' (Oxford English Dictionary [Online] 2021). As well as clarifying what we want to achieve, we also include in this category related questions of scope—defining the boundaries of the scenario exercise in terms of both content and process. We will consider some of these elements (at the level of objective-setting) shortly. Our use of the term *approach* needs more explanation because, in the context of Antarctic geopolitics, we believe that decisions made at the level of approach-setting are crucial—and at risk of being overlooked. An approach, then, is '[a] way of considering or handling something, esp. a problem' (Oxford English Dictionary [Online] 2021). We have chosen this term because it encompasses two distinct but related notions: *fundamental assumptions* and *methodological directions*.

Different fundamental assumptions impact the shape and trajectory of our scenarios. For this reason, it is important to raise awareness—and, if possible, secure agreement—among participants before we start. These assumptions may be about the nature of reality (ontology); about the nature of our knowledge (epistemology); or assumptions about our values (axiology). To illustrate, the 'deductive-normative' approach (see below) is grounded in a worldview that the future is not predetermined. On this view, actors have agency to actively shape their future, and so scenarios can be used to chart a course to a desired goal. These assumptions then guide the broad methodological direction that we select: in this case, we would scan present conditions and then build scenario narratives toward a desired future (Amer et al. 2013; Godet 1986; Godet and Roubelat 1996; Spaniol and Rowland 2018; Durance and Godet 2010). Another fundamental assumption is about the extent to which the core subject matter of our scenarios can be represented in numerically represented data and mathematically expressed logics. Will a quantitative approach (see below) yield gibberish, or can it navigate complexity in ways beyond the reach of human thought?

For us, questions of approach are more wide-ranging and of a higher order of abstraction than is conveyed by the term *technique,* which is 'a particular way of carrying out an experiment, procedure, or task, esp. in a scientific discipline or a craft; a technical or scientific method' (Oxford English Dictionary [Online] 2021). An example of a technique is 'structured scenario fusion', which involves technical procedures to systematically integrate multiple, independently developed scenarios (Barker 2014).

In this section, we have proposed and defined three layers of decision-making for a scenario planning process: strategic *objectives*; general *approaches*; and specific *techniques*. We suggest that an organisation sponsoring a scenario exercise should be conscious of the decisions it is making at each of these three levels. Why? Because these decisions can have profound impacts for the rigour, robustness, and trajectories of the scenarios being created. We will unpack each level further in this chapter.

4.3 Framing Before Reframing

> Get the question right, [and] then choose the method. (Berlin and Solow 2009, cited in Nesbit et al. (2011), p. i14)

Getting the question right means clarifying objectives, defining the scope to be covered, and identifying the types of issue to be considered. It means resisting the temptation to just 'get going' with developing the scenarios themselves and, instead, investing time at the front end of the process in some initial research, consultation, and analysis. Why? Because only then will we be in a position to make informed choices about the most appropriate scenario approaches and techniques for the topic at hand. We have found the OSPA to be particularly useful as a guide on how to 'get the question right'. While there are other excellent introductions to scenario theory and practice (for example, Schwartz 1996), the OSPA stands out for its clear articulation of the conscious choices we need to make when planning a scenario process—at three increasingly granular levels. Under this model, at the first level we define our objectives; at the second level, we clarify our methodological assumptions as we decide upon an approach; at the third level, we select our specific scenario technique/s.

While these are essential preparatory steps, it is important to avoid being so bound by them that we cannot be flexible in our objectives, approaches, and techniques as new insights arrive during the scenario process. This is not to suggest wholesale abandonment of scope and objectives. However, at some point, in an ideal scenario process, we will break at least some of the mould we have cast. We will discard our initial framing of issues in favour of what Ramírez and Wilkinson refer to as reframed and reperceived understandings (Ramírez and Wilkinson 2016). This is the sought-after 'aha' moment we came across in Chap. 3 (Schwartz 1996; Van der Heijden

1996; Wack 1985a). We may even come to the realisation that we have got the initial question 'wrong' and, as a result, need to modify our approach. This is likely to test the skill of the facilitator, the engagement of participants, and the patience of the sponsoring agency.

4.4 Level 1—Define Your Strategic Objectives

A first step is to set the key parameters for the proposed scenario process (Ramírez and Wilkinson 2014). Specifically, we seek to clarify our 'purpose, use and authorizing environment' (Ramírez and Wilkinson 2016, p. 120). At Level 1, Ramírez and Wilkinson suggest we ask the following questions.

- What are our goals?
- Whose understanding is being targeted?
- How will we use the end-products of the scenario process?
- What constraints are imposed by the organisation/s commissioning the process, and the environment in which they operate?
- What are the impacts and considerations if multiple organisations are involved?
- How will the scenarios (and the scenario-building process) be communicated to others who were not directly involved in their creation, so that they are engaged effectively?
- What are the organisational requirements for confidentiality?
- If one of the goals is to reframe understanding, about what, and how, might that reperception be stimulated?
- What resources will be made available?
- Who is to be included (and in what ways)?
- How will the process be governed?
- What organisational characteristics (objectives, cultures, personalities) pose risks to the success of the scenario process? (Ramírez and Wilkinson 2016, pp. 120–146).

The reader is referred to Ramírez and Wilkinson (2016), who examine in detail the many considerations and decisions to be made at Level 1. These decisions will guide the choices we make at subsequent levels about our general approach and specific techniques.

4.5 Level 2—Identify Your General Approach/es

Once an organisation has identified its objectives (and secured executive endorsement for these), the next step is to select an approach or approaches to achieve them. Any scenario approach applied to a geopolitical space needs to simplify the understanding of the space concerned. No single approach can deliver a comprehensive picture.

As Amer and colleagues note, 'Research … indicates that integration of multiple approaches in scenario building process guarantees more robust scenarios' (Amer et al. 2013, p. 30). To achieve this, we can offset the shortcomings of individual approaches with a judicious mix of several. A blend of approaches is likely to deliver a process and outputs that are more rigorous and robust (Kaufmann 2016).

At Level 2, Ramírez and Wilkinson also suggest we need to ask questions such as: (1) what are our methodological assumptions and epistemological principles; and, (2) how do these assumptions and principles impact the practical construction of the scenario intervention? (Ramírez and Wilkinson 2016). At this point, we anticipate that many time-pressed executives and project managers will lose patience. In our experience, it is quite likely that neither the sponsor's board nor many of its employees will have time or interest to consider abstract questions of epistemology and methodology—beyond the most practical considerations. To this, we make two observations. The first is that putting time and effort into prudent assessments at this level will yield a solid return on investment in the form of scenario approaches that are aligned properly with the sponsor's objectives, and in processes that are plausible, rigorous, and robust. The second observation is that much of the spade work done at this level is likely to be a thankless task, with the outcomes perhaps summarised in only a few sentences for executive and external audiences. We suggest this is best handled by a scenario consultant working in close consultation with a small group from the sponsoring organisation and potentially other stakeholders—what Ramírez and Wilkinson refer to as the 'core team' (Ramírez and Wilkinson 2016, p. 126).

4.6 A Scenario Toolbox

At this point, we develop the concept of a 'scenario toolbox'. This is an adaptation of Wright and colleagues' notion of a strategy toolbox (Wright et al. 2013). The terms 'scenario toolbox' and 'scenario toolkit' are used extensively in the scenario literature. A useful toolbox has compartments, where items with similar purpose or features are grouped together. Our survey of the scenario literature has identified at least 80 distinct but overlapping 'ways of doing' scenarios. Together, these ways of doing span all three levels of objectives, approaches, and techniques. There are many possible systems for classifying them (Wright et al. 2013; Huss 1988; Amer et al. 2013; Zentner 1982; Konno et al. 2014a; Chermack 2018), and we will not reprise those typologies here. However, certain common groupings emerge. As a sense-making device, but not a formal taxonomy, the following scenario toolbox provides a crude heuristic that might be used for identifying which of the broad approaches are most suitable for a given set of objectives. These compartments are for explanatory purposes and are not discrete categories. We have found many overlaps and blended approaches, and a specific technique may belong in several compartments at once.

For example, we can find instances where inductive approaches are normative (Kharrazi and Kakuwa 2017) rather than descriptive, and where intuitive logics approaches are inductive (Ramírez and Wilkinson 2014; Bradfield et al. 2005;

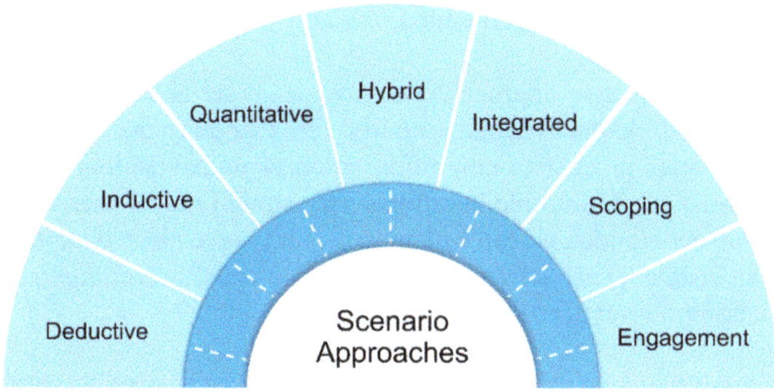

Fig. 4.1 Broad scenario approaches

Leseure 2019) rather than deductive. Intuitive logics (IL) may be approached in other ways also (Ramírez and Wilkinson 2014, 2016; Leseure 2019). The compartments of this toolbox are laid out in Fig. 4.1.

4.7 Deductive Approaches: Intuitive, Archetypal, and Normative

How is *deductive* reasoning used in scenarios? With deductive reasoning, certain conclusions follow as a logical consequence of a given set of premisses (Marcus 1995). And a *premiss* is simply 'a statement or proposition from which another is inferred or follows as a conclusion' (Oxford English Dictionary [Online] 2021). A *deductive* approach to scenarios involves constructing a set of plausible futures, each of which is a logical consequence of a given set of premisses, which are identified at the beginning of the process. Under this approach, we ask two questions: (1) what plausible future states could occur in the future if we apply certain premisses; and, (2) what plausible sequence of events could unfold that would take us from our present state to those future states? Later in this chapter, when describing the IL technique, we will describe these premisses as *driving forces*. Of course, each of these premisses has its own set of underlying premisses! It is a matter of judgement as to how far back in the logic chain we need to go to be satisfied we have a sufficient grasp of the key influences impacting our driving forces.

Under the IL technique (actually a group of techniques), participants construct narratives to describe plausible chains of causation for events between now and a small number of contrasting future states. Commonly, different narratives are created for four distinct end states, whose core characteristics are defined deductively from combinations of critical uncertainties (the premisses). Thus, the core characteristics of these future states are set, in broad terms, at the beginning (Ramírez and Wilkinson 2016; Huss 1988; Amer et al. 2013; Konno et al. 2014a). These deductive approaches can be contrasted with more open-ended, *inductive* approaches, which we explore below.

Importantly, a deductive approach does not mean 'set and forget'. With skilled facilitation, we will know when to modify (or abandon) our original scenario premisses and logics in light of new information or understanding. We may need to run scenarios through multiple iterations. This should not be seen as a failure of process. As van der Heijden notes, 'Learning can only take place if experience deviates from plan in an unexpected way. If everything happens according to expectation there is no learning' (van der Heijden 1996, p. 44).

4.8 Deductive (Intuitive) Approaches

How is *intuitive* thinking used in scenarios? Intuition is often touched on in the scenario literature but is seldom defined. With intuitive thinking, we have a direct relation 'between the mind and something abstract' and which involves 'certain truths regarded as not accessible to investigation through the senses or calculation' (Lacey 1995, p. 415). Mercer sees intuition in scenario processes as the 'ability to make sense of complex patterns of 'soft' data which more rigorous analysis would be unable to handle' (Mercer 1995, p. 84). For Gambelli and colleagues, intuition in scenarios encompasses opinions, initial impressions, experiential thinking, and emotional thinking (Gambelli et al. 2010). For Ramírez and Wilkinson, intuition—which is harnessed to achieve plausibility—entails a process of 'disciplined imagination', in which narratives are co-created iteratively with other learners, who engage with each other's mental models, as well as with the products of analytical research (Ramírez and Wilkinson 2016, p. 66–67). As we saw in Chap. 3, *imagination* is a companion property to structured thinking in scenarios. It is 'the power of the mind to consider things which are not present to the senses', often through the recombination of known elements, using mental imagery (Martin 1995, p. 395). In our experience of risk-averse organisational cultures, many people are reluctant to be imaginative in front of their peers. However, in scenarios practice, imagination is vital if we are to challenge assumptions and transcend incremental thinking. For this reason, sponsoring organisations need to 'give permission' for scenario participants to let their hair down and exercise their (disciplined) imaginations.

Of course, there are methodological challenges in using intuition in scenarios. Intuitive thinking can be hard to render systematically (Schoemaker 1993; Ramírez and Wilkinson 2014), although further down we will see that hybrid approaches can systematically capture at least some aspects of qualitative approaches. At a deeper level, Paya suggests that the products of intuition do not constitute objective, generalisable 'knowledge' in the form of something that is 'publicly accessible and assessable' (Paya 2018, p. 105). This calls into question the relative status of intuitive insights, when compared to other forms of data and analysis used within a scenario process. Ramírez and Wilkinson note Kahneman and Klein's observation that 'professional intuition is sometimes marvellous and sometimes flawed' and that, in practice, it is difficult to 'separate true intuitive skill from overconfident and biased impressions' (Ramírez and Wilkinson 2014, p. 256).

We agree with Ramírez and Wilkinson, who suggest that the epistemological status of intuition in scenarios is problematic only if we expect our scenarios to produce standard prediction extrapolated from past knowledge, rather than using intuition as a means of reframing what is *plausible* (Ramírez and Wilkinson 2014). Without the disruptive role of intuition, the 'aha' moment, and the challenging of existing assumptions, the reframing of perspectives might never occur. For these reasons, we believe that intuitive thinking plays an important complementary role to more structured forms of knowledge and analysis in scenario production (Ramírez and Wilkinson 2014).

4.9 Deductive (Archetype) Approaches

Archetype scenarios use deductive reasoning to ask: what plausible sequence of events could unfold that would take us from our present state to certain archetypal future states? An *archetype* is an original, assumed or ideal pattern or model; in Jungian terms, it represents a 'pervasive idea, image, or symbol that forms part of the collective unconscious' (Oxford English Dictionary [Online] 2021). In other words, we carry these archetypes around with us: they influence our choices, but we are not always fully aware of them. To illustrate, the most basic scenario archetypes are *utopia* and *dystopia* (Wack 1985b). In the context of Antarctica, 'utopian framings' (Hemmings 2013) are often constructed in the absence of a resident human society (Leane 2013) and have tended to centre around permanent security of environmental values and 'saving' Antarctica from various exploitative threats (Hemmings 2013).

Being so easily understood, the utopian-dystopian dichotomy is a tempting approach for the scenario planner. However, it comes with a challenge. Participants may disagree over what premisses constitute utopia (to be aimed for) or dystopia (to be avoided). For this reason, there is merit in considering some alternative archetype models, such as Bezold's four archetypes (Bezold 2009); Dator's four generic images of the future (Dator 2009); Wack's archetypal responses (Wack 1985b); and Young's patterns of institutional dynamics (Young 2010). With skilled facilitation, contrasting worldviews may provoke necessary conversations and allow submerged assumptions to come the surface. Strongly held but conflicting visions need not derail a scenario exercise: if time allows, participants can construct more scenarios to reflect the divergent perspectives. A far greater danger to a successful outcome is posed by inflexible 'right/wrong' thinking.

4.10 Deductive (Normative) Approaches

Deductive (normative) approaches ask: what plausible sequence of events could unfold that would take us from our present state to our *desired* future state (Durance and Godet 2010)? These are goal-directed approaches, built on the premiss that the future is not predetermined but can be actively shaped. A deductive progression is built from in-depth scanning and analyses of the present base and external context, toward desired future images (Amer et al. 2013; Godet 1986; Godet and Roubelat 1996; Spaniol and Rowland 2018; Durance and Godet 2010). A pre-eminent example of deductive (normative) approach is *la prospective*, a mindset that rejects determinism and instead promotes the agency of actors to shape their future. By agency, we mean having both the freedom and capacity to act independently.[3] The core principle of *la prospective* is that a rigorous scenario process can lead to good planning, and good planning can spur us to action in the present leading to a preferred outcome in the future (Spaniol and Rowland 2018). Crucially, Godet and Roubelat warn that such a process is technically demanding and time consuming and should take around a year. We concur. We argue throughout this book that, regardless of approach, committing time and resources to scenario development on this scale represents a very small fraction of the national investments directed to Antarctic programs.

4.11 Inductive Approaches

How is *inductive* reasoning used in scenarios? Before continuing, to better understand what we mean by inductive reasoning, it may be useful to recap what we have said about *deductive* reasoning. Using deduction, certain conclusions follow as a logical consequence of a given set of premisses (Marcus 1995). These premisses are generally identified at the beginning of the process from key driving forces that operate in a given field, or from archetypal images of alternative futures. In short, with a deductive approach we *work backwards* from a defined set of premisses, noting that our premisses may need to evolve during a scenario process.

In contrast, *inductive* reasoning is '[t]he process of inferring a general law or principle from the observation of particular instances' (Oxford English Dictionary [Online] 2021). Here, our conclusions are not deduced logically from a given set of premisses, but 'rendered plausible in the light of them' (Cohen and Honderich 1995, p. 405). An inductive approach to scenarios, then, involves non-prescriptive, open-ended or 'bottom-up' (O'Mahony 2014) methods that take present events [or

[3] The *la prospective* approach, named in 1959 by philosopher Gaston Berger, holds that 'the future is not part of a "predetermined temporal continuity" but something which is to be created and which can be "consciously modeled to be humanly beneficial"' (Schwartz 1996, p. 802). As Bradfield and colleagues note, 'The primary objective of the Prospectives centre was to formulate an acceptable scenario-based methodology for developing positive images or "normative scenarios" of the future and to lead these images into the political arena where they could serve as a guiding vision to policy makers and the nation, providing a basis for action' (Schwartz 1996, p. 802).

possibly a single significant event (Konno et al. 2014a)] or variables as the point of embarkation, then *work forward* in time to see what chains of events and future states might unfold (Konno et al. 2014a). The inductive approach should not be confused with a purely intuitive, 'anything goes' mindset. In some cases, rigorous processes have been developed to support inductive reasoning in scenarios, incorporating data and information analytics, game theory analyses, and expert validation of causal plausibility (Sharma and Yang 2015).

Induction enables participants to work more freely with a larger number of driving forces than is possible under the deductive (intuitive logics) approach (Talberg et al. 2018; Freeth and Drimie 2016; Schnaars 1987). In Chap. 5, we will argue that, in a complex field such as Antarctic geopolitics, this is a major benefit. Also, when compared with deductive methods, inductive approaches may be quite liberating for participants [who may then produce more surprising outcomes (Wodak and Neale 2015; Nygrén 2019)] because they are not confined to constructing scenarios within given sets of premisses. However, inductive processes tend to be time consuming, they are 'often more difficult to arrange' (Nygrén 2019, p. 33; Wodak and Neale 2015), and without the corralling effect of premisses, it can be hard to justify why one hypothetical turn of events is selected within a scenario, rather than another (Konno et al. 2014a; Lord et al. 2016).

4.12 Quantitative approaches

As Lindgren and Bandhold put it, quantitative approaches provide 'output data in numbers' (Lindgren and Bandhold 2009, p. 186). Quantitative approaches encompass a large suite of methods that use numerically represented data and mathematically expressed logics to: model scale, intensity, probability and sensitivity; identify structures and relationships; integrate data and variables; extrapolate trends; simulate impacts of interventions; exclude mutually incompatible combinations of factors; and process larger volumes of data, variables and relationships than can be handled by qualitative approaches alone (Wright et al. 2013; Huss 1988; Amer et al. 2013; Wodak and Neale 2015; Nygrén 2019; Lord et al. 2016; Bradfield et al. 2016; Mack 2010; Halim et al. 2016).

Scenarios are not needed when *all* the key variables in a given field can be represented mathematically and when quantitative models can provide a replicable means to test how a system will evolve under different conditions (O'Mahony 2014). As we shall see, the field of Antarctic geopolitics is neither entirely reducible to mathematical representation, nor is it sufficiently stable to allow replicable modelling of its variables over time. For this reason, we argue that, here, the use of scenarios is entirely appropriate. We will also suggest that quantitative approaches have an important role to play.

Drawing on Lindgren and Bandhold, we note at least three applications of quantitative approaches in scenarios. The first use is to construct future trajectories of phenomena, on the assumption that they can be captured and simulated mathematically through techniques such as regression analysis and systems models (Lindgren and Bandhold 2009). Regression analysis focuses initially on identifying correlations, after which causal inferences may be drawn (with caution) from further statistical tests. The second application for quantitative approaches is to provide 'solid facts and figures' (Lindgren and Bandhold 2009, p. 186), as data inputs to inform qualitative reasoning in scenario development. The third application is within hybrid quantitative/qualitative approaches (Lindgren and Bandhold 2009), where quantitative analysis is used to augment qualitative reasoning (see below). In the following chapters, we will suggest that the judicious use of quantitative techniques is an under-developed area for geopolitics in general and for Antarctic geopolitics in particular.

The main advantage of quantitative approaches for scenarios is their capacity to systematically capture, process, and manipulate certain types of information on a scale and to a level of complexity that humans cannot match. Here, we make three points. The first is that this processing capability is useful when we need to handle large quantities of relevant data. In Chaps. 5 and 6, we argue that if our scenarios on Antarctic geopolitics are to be rigorous and robust (Amer et al. 2013), then they will need to process a considerable variety and volume of information—at least some of which can be quantified. For this reason alone, quantitative approaches should be actively considered. The second point is that quantitative approaches also provide the ability to project historical trends forward into the future using statistical and probabilistic techniques (Amer et al. 2013) on a scale and to a level of complexity that cannot be achieved by human intuition alone. The third (and perhaps most interesting) contribution of quantitative approaches is their potential to contribute to those 'aha' moments of unexpected insight.

There are also disadvantages to the quantitative approach. First, as we saw in Chap. 3, due to the convergence of complexity, uncertainty and the passage of time, the accuracy of extrapolative techniques will diminish the further ahead we look. For this reason, quantitative approaches are best suited to scenarios that are narrow in their subject matter (Amer et al. 2013) and short in their time horizon (Amer et al. 2013; Lindgren and Bandhold 2009). Second, quantitative approaches have limited ability to handle certain types of data and reasoning (Lindgren and Bandhold 2009). These include matters of governance, social and cultural drivers (O'Mahony 2014), as well as attitudes, politics, and some aspects of economics (Amer et al. 2013). Writing in 2001, Martelli noted the inability of computational methods to account for 'soft concepts' such as human values and experience (Martelli 2001, para 41). However, recent advances in artificial intelligence may be narrowing the gap between hard and soft concepts. Third, in Chap. 3 we suggested that scenarios operate in the zone of plausibility, and that one of the key properties of that zone is complexity. However, many complex systems are not readily reduced, described, or modelled in purely quantitative terms (Ritchey 2011). The fourth disadvantage is participant disengagement from quantitative approaches. We call this *black box disempowerment,* with the 'black box' referring to those aspects of a quantitative process that are not accessible (and able to be critically scrutinised) by the intelligent layperson.

Moreover, under a quantitative approach, the scenario sponsor and participants may be forced to rely heavily on external experts. As Bradfield and colleagues note, in highly quantitative techniques, 'external experts unavoidably play a dominant role as consultants in designing and then carrying out the scenario exercise' (Bradfield et al. 2005, p. 809). For these reasons, we follow Martelli in recommending that quantitative approaches be restricted to the 'bare, essential minimum' (Martelli 2001, para 40).

4.13 Hybrid approaches

> [S]cenario planning approaches comprising of a combination of qualitative and quantitative techniques are better and can result in … robust scenarios. (Amer et al. 2013, p. 38)

Within the scenario literature, 'hybrid' generally refers to a specific group of techniques for blending quantitative and qualitative approaches. These tend to be sophisticated processes in which quantitative models are used to structure and analyse causation and patterns, drawing on qualitative inputs (O'Mahony 2014; Nygrén 2019; Lord et al. 2016; Ritchey 2011; Varho and Tapio 2013; Tapio et al. 2017). In other words, qualitative data form significant inputs within a primarily quantitative process. These qualitative inputs include what Raskin and colleagues identify as the 'ultimate drivers' or 'conditioning framework' of a system—such as power, culture and values (O'Mahony 2014, p. 44). These ultimate drivers can have a profound bearing on quantifiable inputs in areas such as demographics and economics (Cairns et al. 2017).

In practice, how does the interface between quantitative and qualitative approaches work? At its simplest, it involves quantifying and then modelling qualitative inputs in terms of their scale and intensity (O'Mahony 2014). At a more sophisticated level, the interface can involve generating qualitative inputs, running them through various analytical processes to develop and constrain alternative scenario outcomes, and then using qualitative judgements to assess the plausibility of the outputs of those processes (Lord et al. 2016). These hybrid approaches have been referred to as the 'frontier of scenario research' (O'Mahony 2014, p. 44).

As a means to circumvent the respective limitations of quantitative and qualitative approaches, hybrid approaches have strong theoretical appeal. For example, later in this chapter, we will see that that under the 'intuitive logics (IL) 2 × 2' qualitative technique, scenario outcomes are constrained because each scenario can only deal with two key factors ('driving forces') at a time. Integrating the IL 2 × 2 scenario logic with structured analysis and quantitative modelling approaches offers the potential to significantly increase the number of driving forces under consideration within a single scenario (O'Mahony 2014; Lord et al. 2016). As O'Mahony notes, there are other benefits:

> The use of the narratives to structure thinking on the alternative development pathways not only makes scenarios more believable but can ensure that they do not become an arbitrary numerical combination of quantitative variables and can make assumptions more explicit.

> In parallel, the quantitative scenarios are complimentary to the qualitative narratives. They are used to check consistency, to provide relevant numerical information, and to 'enrich' the qualitative scenarios by showing trends and dynamics not anticipated by the storylines. (O'Mahony 2014, p. 44)

However, as Nygrén observes, in practice, combining these two approaches is 'laborious and not an easy task' (Nygrén 2019, p. 32). In the field of scenarios for carbon emissions, for example, O'Mahony observes a dualism between quantitative and qualitative approaches, with 'two largely non-overlapping streams of inquiry' (O'Mahony 2014, p. 44).

There is also a second sense in which quantitative and qualitative approaches can be combined. This is where quantitative data form significant inputs within a process that is primarily based on qualitative scenario reasoning. We will refer to these as *informal hybrid* approaches. Here, decisions about how the quantitative data will be used and what weight to give it will be qualitative judgements made by the scenario facilitator, core team and participants. This approach makes use of quantitative *data* while eliminating the risk of black box disempowerment—and is much simpler to administer.

4.14 Integrated approaches

As their name suggests, integrated approaches attempt the ambitious (but arguably necessary) task of *integrating* all relevant inputs into the scenario analysis. The task is ambitious because in any field there will be a lot to cover, causal relationships will be complex, and it will be all too easy for the process to drown in detail and go nowhere. However, the task is arguably necessary because a non-integrated approach will, by definition, be incomplete—possibly missing crucial factors at work in a given field. Integrated approaches are designed to tackle complexity and minimise the artificial segregation of related factors by: integrating data drawn from different levels and sectors (Nicosia 2011); analysing the causal impacts of converging factors (Schoemaker 1993); identifying cross-impacts (Huss 1988; Amer et al. 2013; Godet and Roubelat 1996); assimilating multiple, concurrent scenarios (Amer et al. 2013; Barker 2014); and modelling the future impacts of multiple policies with conflicting objectives (Bradfield et al. 2016).

The integration effort can span different scales (Young 2002) and types of subject matter. These can blend *qualitative* and *quantitative* data and analysis; assess causal impacts when *various factors* interact; synthesise information across *different scales* (temporal and spatial) (Young 2002); amalgamate information drawn from *disparate disciplines*; and integrate the results of *independently developed scenarios*. As Shoemaker asserts, 'The efficacy of scenario planning seems to hinge on examining a confluence of factors (instead of just single contingencies)' (Schoemaker 1993, p. 209). For example, cross-impact analysis proceeds on the basis that it is not realistic to forecast a particular occurrence in isolation of other key impacting events (Amer et al. 2013). Instead, we need to assess the causal implications of those impacting events for our scenario narratives (Huss 1988). One way of doing this is to calculate the likely impact that one event has on the probability of other key events occurring (Huss 1988). Managing a process of this kind is complex and demands skilled facilitation to avoid bogging the process down by examining too many factors at once.

Another issue tackled by integrated approaches is how to deal with *multiple independent scenarios.* For example, in the security sector, there is a growing interrelatedness of issues and a need for agencies to plan for 'more, smaller, and interconnected crises occurring simultaneously' (Barker 2014, p. 1726). However, the difficulty of linking qualitative and qualitative approaches has proven to be 'a major obstacle' in progressing integrated scenarios [Kok and van Delden (2009) cited in Amer et al. (2013), p. 30]. Nonetheless, we suggest that some form of basic integration is desirable, if our scenarios are to come close to matching the level of complexity of the field we are examining.

4.15 Scoping approaches

Scoping scenarios probe for specific perspectives and data, in terms of levels of aggregation, sectoral focus, timeframes, organisational priorities and the probability of outcomes. A rich array of options is available (Table 4.1), and the optimal choice will depend on the sponsoring organisation's objectives. At a minimum, we recommend some combination of macro- *and* micro-level perspectives. This will ensure that scenarios 'cover the ground' and capture most key driving forces—some of which may operate at one of these levels, but not the other (Kok 2016).

4.16 Engagement approaches

These techniques are designed variously to foster participant engagement with the scenario process; strengthen the calibre of inputs; and challenge prevailing perspectives. Collaborative/participative/participatory approaches provide a range of benefits, both to the success of the scenarios project itself and to the participants (Ritchey 2011). Engagement approaches can elicit expertise; multidisciplinary contributions; collaboration; participation; shared understanding; creativity; unorthodoxy; provocation; debate; questioning of assumptions and mindsets; avoidance of groupthink; self-awareness; experiential awareness of future states; immersion; personal connection; awareness of winners, losers, and power relationships; heightened awareness of personal agency to make changes towards a desired future; inclusion of disempowered voices; strategic conversations; and an orientation to futures thinking (Wright et al. 2013; Ramírez et al. 2015; Cairns et al. 2013; Wodak and Neale 2015; Nygrén 2019; Riddell et al. 2018; Fink et al. 2010; Totin et al. 2018; Baena 2017; Meissner et al. 2017).

Table 4.1 Scoping options for scenarios

Focus	Description	References
Macroscope	Scenarios that focus on 'the wider, complex inter-connected system in which an organization exists'	Amer et al. (2013, p. 25)
Global/First generation/Learning	Initial macro-view scenarios that are wide-focused to capture the full context and all key variables	Amer et al. (2013), Bradfield et al. (2016), Chermack and Coons (2015)
Global analytical	Scenarios that strive for a holistic integration of information between scale levels, and across disciplinary sectors	Nicosia (2011)
Multi-disciplinary	Scenarios with a holistic framework to integrate central ideas, capturing views of 'scientists, engineers, psychologists and even mystics'	Amer et al. (2013, p. 26)
Specific/Second generation/Decision	Subsequent, micro-view scenarios, which are narrow-focused; undertaken once global scenarios have been developed	Amer et al. (2013), Bradfield et al. (2016), Chermack and Coons (2015)
Internal	Scenarios that focus on endogenous responses within an organisation to exogenous changes that arise in scenarios	Tapinos (2012), Wright et al. (2013)
Baseline/Business-as-usual	Scenarios that model futures, assuming no further policy interventions or mitigating measures are made	O'Mahony (2014), Hasegawa et al. (2018)
Policy	Scenarios that model futures under certain policy interventions and mitigating measures	O'Mahony (2014)
Policy-oriented/Exploratory	Exploratory scenarios, which are framed and developed through the lens of policy effectiveness, using participatory processes	Riddell et al. (2018)

(continued)

Table 4.1 (continued)

Focus	Description	References
Goal-based	Scenarios in which participants pursue an objective amid multiple competing goals, in a simulated scenario environment	Bell and Marks (2016)
Organisation-based	Scenarios that model organisational responses to exogenous changes as they unfold within scenarios; institutional dynamics; available policy instruments/levers	Wright et al. (2013), Amer et al. (2013)
Time-based	Scenarios that sequence analysis of the future into stages of development	Barker (2014)
Strategic early warning	Scenarios that prepare for dynamic change by identifying and evaluating weak signals	Fink et al. (2010)
Analogue	Future scenarios that are constructed based on corresponding circumstances from other spatial regions or time periods	Mearns et al. (2001)
Critical	Scenarios that focus on impacts of scenario decisions for those not directly involved/less visible; perspectives of ethics, power relations; winners and losers	Wright et al. (2013)
Transformative	Scenarios that focus on situations that are unacceptable, and where actors cannot change things directly, or act within existing alliances	Freeth and Drimie (2016), Kahane (2012), Totin et al. (2018)
La prospective: desirable	Scenarios that are possible and desired, but not necessarily realisable	Godet and Roubelat (1996)
La prospective: normative	Scenarios that analyse the current base and external context, then simulate a causal progression toward desired future images	Amer et al. (2013), Godet (1986), Godet and Roubelat (1996), Spaniol and Rowland (2018), Durance and Godet (2010)
La prospective: possible	Scenarios that contemplate 'everything that can be imagined'	Godet and Roubelat (1996, p. 167)

(continued)

Table 4.1 (continued)

Focus	Description	References
La prospective: realisable	Scenarios that contemplate 'all that is possible, [but] taking account of constraints'	Godet and Roubelat (1996, p. 167)
La prospective: reference	Scenarios that identify the most probable causal path for the relevant environment and specific variables under study	Godet (1986)
La prospective: trend-based	Scenarios that extrapolate current trends into the future, 'at all points where choices are to be made'	Godet and Roubelat (1996, p. 167), Billen et al. (2018)

Complex problems, in particular, benefit from engaging multiple perspectives (Nygrén 2019). In addition, some engagement techniques may also improve predictive accuracy. For example, using historic case studies, scenario participants who engaged in intuitive *role playing* achieved high levels of predictive accuracy in responses to conflict situations (Green and Armstrong (2011), cited in Wright et al. (2013)). Participant engagement is also essential if we are to foster commitment to the outcomes of the scenario process from organisational members and other stakeholders. Ideally, the scenario environment will be one of non-threatening exploration of alternative futures (Wright et al. 2013), for 'pseudo-experimentation' (Wright et al. 2013), and where it is safe for participants to take risks, to 'think the unthinkable' (Kahn and Wiener (1967) cited in Wright et al. (2013), p. 3). At times, however, significant personal courage is needed to challenge and be challenged about our prevailing assumptions and aspirations for the future. Fault lines within organisations and between stakeholders may be forced to the surface or magnified.

As Kahane notes, some problematic issues can become polarised, confused or 'stuck' (Kahane 2012, p. 18). These issues tend to be:

- 'unacceptable, unstable, or unsustainable';
- complex, interdependent, and unpredictable; and
- incapable of being transformed directly by an actor (or group of like-minded actors) when operating unilaterally (Kahane 2012, p. 16).

In these cases, it may not be effective to plough on with preconceived strategies that can only be adapted at the margins. Instead, an engagement approach using 'participatory' or 'transformative' (Nygrén 2019; Kahane 2012) techniques may be appropriate. This involves reaching across divides and transforming stakeholder understandings, relationships, intentions, and actions (Kahane 2012). For some of the more contentious topics in Antarctic geopolitics, we suggest there might be value in conducting scenarios using both traditional *and* engagement approaches—perhaps in an iterative cycle.

4.17 Level 3—Select Your Specific Technique/s

So far, we have sketched out the field of scenario objectives and approaches in broad strokes. We have now arrived at the level of specific techniques. For reasons of space, we cannot detail all the available scenario techniques here. However, the reader may wish to explore the works referenced under the various approaches listed above.

To make scenario practice more concrete, we will illustrate one prominent technique in depth.

4.18 The 'Intuitive Logics 2 × 2' Technique

This section elaborates one form of the intuitive logics (IL) process pioneered in the late 1960s by Pierre Wack and others at Royal Dutch Shell. The specific technique we examine is the *standard IL 2 × 2 process*. This was selected for illustration and deeper analysis because it is a relatively straightforward deductive technique that is often applied in geopolitical scenarios.[4,5] In order to explore the IL approach, we provide (below) a worked example to illustrate a number of its core concepts, which inform the discussion that follows.

The example we have chosen unfolds over nine steps. As van Efferink notes, for scenarios to be credible, the preliminary steps (1–6) typically require considerable data collection, research and analysis *before* scenario drafting begins.[6]

4.18.1 Step 1: Identify Your Scenario Standpoint

Effective scenarios (of the type illustrated here) are conducted from one perspective. This is the single standpoint of the actor (or group of actors) whose interests, threats, strategies, and tactical decisions we will 'stress test' as our scenarios unfold. The perspective we select may be that of a single actor, or a coherent coalition of interests. It is this perspective that is used to identify a coherent set of critical uncertainties (see below) that are relevant to that actor, or group of actors. Our scenario standpoint should be clear to all participants if we have already defined our strategic objectives at Level 1 (see above). If we take a classical approach to geopolitics, then we would

[4] The authors thank Dr. Leonhardt van Efferink (Geopolitics and Country Risk, Rotterdam) and the Maastricht University Summer School for the many strategic insights and practical tips about the IL method, provided during the Maastricht University Summer School, 'Geopolitical Scenario Planning' (MS0040), Maastricht, The Netherlands, 5–9 August 2019. The authors have adapted this knowledge (with permission) for use in this section. However, any errors remain with the authors.

[5] For a detailed treatment of this method and its many variants, see Ramírez et al. (2015). Other excellent accounts include Schwartz (1996), Wright et al. (2013), (Schoemaker 1993), and Ramírez and Wilkinson (2016).

[6] van Efferink, 'Geopolitical Scenario Planning'.

typically adopt the stance of a single state, or a group of states with similar strategic objectives on a topic. In this exercise, the perspective adopted is that of 'Country A'.

There are two provisos to our need for a single standpoint. First, if our scenarios are to be plausible, rigorous, and robust, then of course other stakeholder perspectives must be examined closely. So, here, we do not ignore other viewpoints. Instead, we simply reflect that a scenario *of this type* cannot be conducted from the diverse perspectives of multiple stakeholders simultaneously. Our second proviso is that some topics may be so problematic and divisive that an *engagement approach*, using participatory or transformative techniques, may be appropriate (see above). In those cases, we suggest that an engagement approach could augment the single standpoint approach used here.

Having selected Country A as our standpoint, we identify its key strategic interests relevant to the topic. This analysis includes constraints on policies and potential threats to those interests and policy strategies (Miller and Siggins 2003), as well as the capabilities of Country A relative to other relevant actors.[7] This analysis considers both domestic and external influences.

4.18.2 Step 2: Set Your Time Horizon

In Chap. 3, we noted that scenarios become useful as we approach the limits of prediction, and these limits are defined by three core factors: *complexity*, *uncertainty*, and *the passage of time*. In a dynamic field such as geopolitics, we arrive at the limits of prediction quite soon. How soon will be influenced by the number and stability of the key variables under investigation (United Nations Industrial Development Organization 2005). Indeed, geopolitical scenarios may prove useful in timeframes of under a year. So, how far into the future should we look? The notion of 'long-range' is elastic and varies with the context (Leigh 2003). For this reason, we suggest that the maximum timeframe for our scenarios is dictated primarily by the planning horizon of the organisation/s sponsoring the scenario exercise. This tends to vary according to whether plans are based around the organisation (often around four years); a subject (around 10 years); a region (around 20 years); or environmental/intergenerational strategies (many decades or more) (Glenn et al. 2010; Miller and Siggins 2003).

4.18.3 Step 3: Identify the Key Driving Forces

Next, to create scenarios relating to a conflict or spatial rivalry, we need to identify the key geopolitical *driving forces* of that conflict or rivalry. We might see these as 'driving' in the sense of '[i]mpelling, setting in motion, actuating' (Oxford English

[7] See Footnote 6.

Dictionary [Online] 2021). As Huss notes, 'scenarios provide a means of differentiating between various interpretations of the interplay of … forces' (Huss 1988, p. 22). In scenarios using deductive reasoning, these are the premisses from which scenario conclusions follow. They are the causal factors or 'trends' (Lindgren and Bandhold 2009) that are important to understanding a conflict or rivalry, and for determining the structure and outcome of the processes. These driving forces are often grouped within the following categories: Politics, Economics, Social, Technological, Environmental, and Legal (PESTEL) (Wright et al. 2013; Liggett et al. 2017).

Driving forces are identified and categorised through the use of planners and analysts, outside consultants, specialised information services, business models, environmental monitoring and scanning systems, general literature about the future (Huss 1988, p. 22), and, importantly, the knowledge and experience of the scenario participants themselves.

4.18.4 Step 4: Distinguish Critical Uncertainties from Predetermined Elements

The next step involves deciding whether the driving forces (which have already been divided into the six PESTEL categories) are either *predetermined elements* or *critical uncertainties*. Predetermined elements are relevant factors that are:

- relatively stable and easy to predict;
- 'what we know we know' (Schwartz 1996), p. 109];
- events that have already happened, or are likely to happen (Amer et al. 2013); and
- *assumptions* about continuity of driving forces from the present into the future.

In the words of Schwartz, these are the slow-changing phenomena, constrained situations, things in the pipeline, and inevitable collisions (Schwartz 1996, pp. 111–112).

In contrast, critical uncertainties (Schwartz 1996; Van der Heijden 1996) are factors that are:

- highly unpredictable aspects of the future;
- unmeasurable in terms of risk, even with the right information;
- likely to have a critical impact on conflict or rivalry outcomes; and
- impossible to influence from the standpoint of the organisation sponsoring the scenario.

Of course, predetermined elements and critical uncertainties can be 'intimately related' (Schwartz 1996, p. 113); their boundaries can be hard to draw, and may change over time. We will return to this in Chap. 7—Antarctic Militarisation: Scenario Analysis.

4.18.5 Step 5: Rank Critical Uncertainties by Importance and Uncertainty

In Step 5, the critical uncertainties are ranked. This ranking takes account of both the *importance* and level of *uncertainty* of a driving force (Schwartz 1996). Here, we are sifting for those driving forces that seem (at the outset) to be the most crucial and the most in doubt (Schwartz 1996). In this context, 'importance' relates to the strategic value of an issue, and its relevance to scenario objectives. A diverse group may disagree on what the top issues are, so it is essential to define the scenario standpoint from the outset (see Step 1). As Wright and colleagues state, in selecting critical uncertainties for use in our scenarios, 'the only rule is that the extremes should be defined creatively to test the limits of thinking, but without breaching the boundaries of possibility and plausibility within the project time horizon' (Wright et al. 2013, p. 6).

4.18.6 Step 6: Select Your General Approach/es

The next step involves an important choice by the analyst to select their general approach (see the Scenario Toolbox, above). For example, we might choose either a *deductive* or *inductive* approach. As we have seen, with a deductive approach we *work backwards* from a defined set of premisses; with an inductive approach we *work forwards* in time to see what chains of events and future states might unfold. Here, we will illustrate a deductive approach because it is generally less time- and resource-intensive for facilitators and participants.[8] And so, using the intuitive IL 2 × 2 technique, we will construct four scenarios whose premisses are defined by the top two critical uncertainties identified at Step 5 (Ramírez et al. 2015; Konno et al. 2014a).

4.18.7 Step 7: Construct the Scenario Matrix

Having adopted the deductive method for this example, we will name the two critical uncertainties selected for our scenarios 'CU 1' and 'CU 2'. Critical uncertainties are selected because they are ranked as most important (Schwartz 1996); they can be defined clearly; they can be recognised—feasibly—at very low and very high values; and they are causally independent of each other. Otherwise, there is too much complexity within the logic of the model (Wright et al. 2013). We will call these four instances, at their low and high values: CU 1 low, CU 1 high; and CU 2 low, CU 2 high.

[8] See Footnote 6.

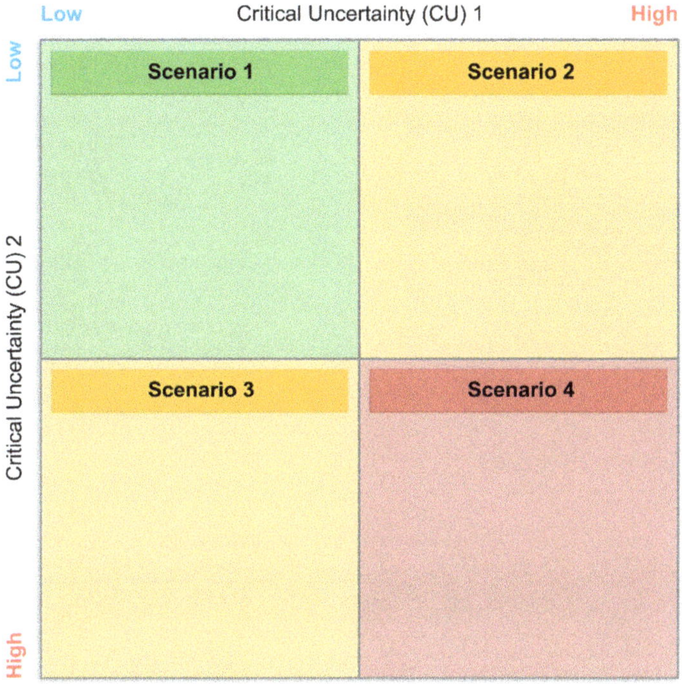

Fig. 4.2 Scenario matrix

We are now in a position to assemble the scenario matrix defined by the two crucial uncertainties selected in earlier steps. This sets out the parameters for our four scenarios (Fig. 4.2).

Here, the scenario is constructed from the viewpoint of one party (in this case Country A).[9] Let us assume that, in the preceding steps, we have identified the following national interests, policies and strategic threats:

- Country A's policy is to uphold the Antarctic Treaty System and its instruments, including the ban on mining under the Madrid Protocol;
- Country A has concerns that unauthorised mining activity undertaken by another country within the territory of Country A's sovereignty claim could be perceived as a de facto challenge to that claim;
- Country B's 'rational use' policy for Antarctica means it is open to commercial mineral extraction *if* economic and logistic conditions become favourable; and
- Country B does not recognise the sovereignty claims of any claimant state under the Antarctic Treaty.

So, here we have a potential geopolitical rivalry over the use of space within Antarctica.

---

[9] See Footnote 6.

Fig. 4.3 Two critical uncertainties

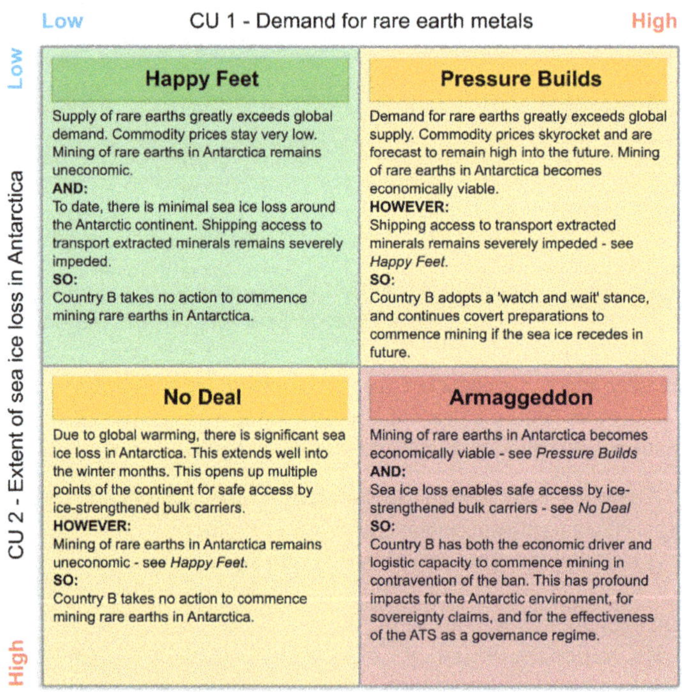

Fig. 4.4 Scenario matrix—worked example

Next, let us assume that we have identified the two critical uncertainties below as the most important and uncertain in this field (Fig. 4.3).

Figure 4.4 illustrates how the interaction of these two critical uncertainties—at high and low values—sets up the scenario matrix and defines the parameters for our four scenarios (Wright et al. 2013). Each scenario is given a brief, evocative title to reflect its fundamental characteristics: this can establish a shorthand vocabulary for subsequent futures-oriented discussions within an organisation.

4.18.8 Step 8: Develop the Scenario Narratives

Having defined the parameters for the scenarios at Step 7, above, the next step is to craft four plausible and internally consistent narratives, drawing on the research materials assembled earlier. As noted, these narratives set out the causal chains that link the end-state of the scenario back to the present. Ideally, the four scenarios should: span the same time horizon; be fundamentally different from each other; have roughly equal likelihood; and have similar strategic value. Also, we will hold the core policy and strategies of Country A constant though each of the scenarios, and across time. This prevents unstable scenarios with too many changing variables. It also enables participants to assess the extent to which *existing* strategy is preparing the organisation for different scenarios, and to identify where that strategy needs to be adjusted.

Within those constraints, the scenarios need to countenance the unthinkable, crazy, impossible, or unknowable, as well as the possibility of a 'black swan' event—something that is unforeseen and rare, but whose antecedents are recognisable after the event (Taleb 2010). The next step is to refine these possibilities so that they fall within the *zone of plausibility* (see Chap. 3). Here, we steer a middle course between the twin hazards of: (a) thinking too small, with incremental extrapolation; and (b) thinking too large, with wild invention untethered to an evidence base. Finally, it is important that the scenarios present the unfolding causal sequence in an engaging and compelling narrative, appropriate to the audience (Wright et al. 2013).

4.18.9 Step 9: 'Wind Tunnel Test' Current Strategy and Prepare Policy Advice

In the final step, the analyst should ask, 'So what?', in the sense of 'What has the IL scenario process shown?' and 'Are the scenarios and scenario space plausible?' This is where we 'wind tunnel test' (Wright et al. 2013; Leseure 2019) Country A's present policies, strategies and capabilities, to see how robust they would be if any of the four scenarios were to eventuate. Ideally, this informs advice to the government of Country A on possible modifications to its current policies, strategies and capabilities.

In practice, how might we undertake a 'wind-tunnel test' of the scenarios? Wright and colleagues propose the following steps, based on the objectives the organisation wishes to achieve in strategic actions, and an assessment of these against alternative strategies:

- Stage 1: Formulate scenarios;
- Stage 2: Formulate the objectives that you wish to achieve in your strategic actions;
- Stage 3: Design alternative strategies;
- Stage 4: For each objective, rank each strategy against each scenario from the best to the worst;
- Stage 5: For each objective, rank all strategy–scenario combinations from best to worst;
- Stage 6: Compute the sum-of-ranks for each strategy and provisionally select the best performing strategy' (Wright et al. 2013, p. 5).

While this ranking of strategy objectives against scenarios may be based on 'rough and ready judgements' (Wright et al. 2013, p. 5), it provides greater rigour to future-proofing a strategy than an informal approach.

4.19 Insights from the Process and Adding Depth: An Alternative to IL 2 × 2—Towards 'IL 5'

In the IL 2 × 2 approach, we explore two critical uncertainties at their low and high values: CU 1 low, CU 1 high; and CU 2 low, CU 2 high. This bifurcation works very neatly to produce well-contrasted outcomes within the logic of the IL 2 × 2 framework. However, *if* our scenarios are to provide targeted, broad-brush prediction, then a middle option may also be needed. Here, we are saying that our strategies need to be somewhat future-proofed against both CU 1 low and CU 1 high, but that, based on available evidence, another option is CU 1 medium.

Here, we are careful to suggest the use of 'medium' in the sense of 'expectation value' in quantum physics, and *not* as the most probable value. The expectation value is the '(probability-weighted) average value of the results' [Cresser 2011, p. 209]. Here, we wish to convey a subtle distinction that positions the medium level of a critical uncertainty as something that is *probabilistically in the middle*, rather than as something that is *more probable*. In other words, we wish to avoid any conclusion that the medium value is more likely. In particular, our scenario planning should not fall into the trap of giving the medium value most of our attention, at the expense of the extremities of high and low. Instead, we are simply offering an additional scenario in which the two critical uncertainties are not met at their extremes. To accommodate this additional, medium value, our intuitive logics method needs to shift slightly from an IL 2 × 2 approach.

Here, we must pause to clarify our terminology. In particular, we wish to distinguish the introduction of a third (medium) value for our two critical uncertainties from the introduction of a third critical uncertainty. Schwartz (2011) proposes that a third critical uncertainty can be introduced into the scenario logic, which converts a two-dimensional matrix into a three-dimensional cube.

Noting Schwartz's concept, Ramírez and Wilkinson refer to this 'volume of possibility' as a '3 × 3' matrix, (Ramírez and Wilkison 2014, p. 257). We suggest, however, that a three-dimensional scenario logic with *three* critical uncertainties, each of which is at *two* levels (high and low) is more accurately described as a '2 × 2 x 2' matrix. Why does this matter? Because here we wish to appropriate the IL 3 × 3 label for a different task: to construct a scenario logic that provides some targeted, broad-brush prediction, with *two* critical uncertainties, each of which is instantiated at *three* levels (high, medium and low)—see Fig. 4.5, overleaf.

Before continuing, we should consider *how many* scenarios we can deal with at once. Too *few* scenarios will not provide a sufficiently diverse set and may lead to participants viewing one scenario as 'good' and the other as 'bad', thus creating a one-dimensional scenario set. A set of three scenarios tends to support participants' zeroing in on one scenario as the 'likely' scenario or as a 'baseline'.

On the other hand, developing a large number of scenarios is resource-intensive, and cognitively overwhelming for participants. Even a low-dimensional morphological analysis can contain hundreds and potentially thousands of plausible scenarios. Practice shows that between four and six scenarios is the best number for qualitative participatory exploration of highly uncertain futures. In part, this is due to the limits of our cognition:

> [P]eople can hold from five to nine information elements for no more than 20 s, and even fewer when the information elements need to be combined, contrasted, or manipulated. (Paas and van Merriënboer 2020, p. 395)

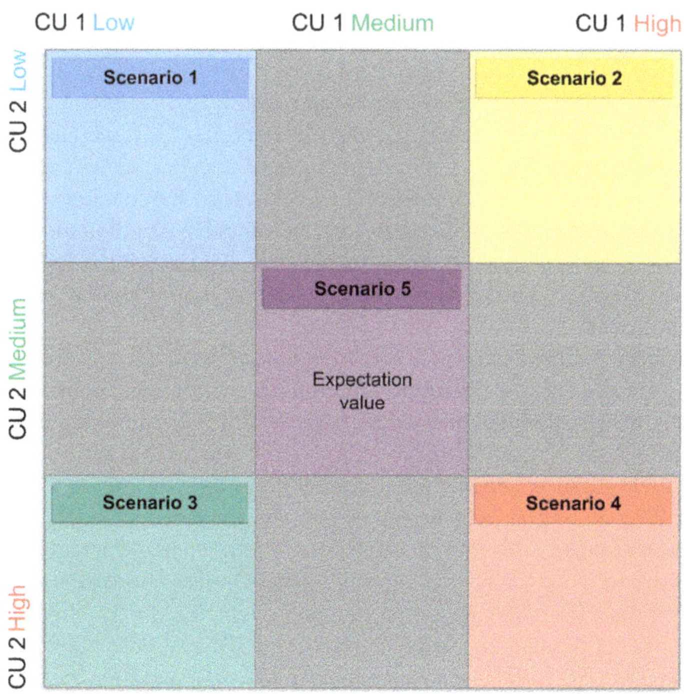

Fig. 4.5 Two critical uncertainties at three levels (high, medium and low)

Confronted with too many scenarios (each representing multi-dimensional narratives on the future), participants will lose sight of the whole and will be less able to sense how the scenarios are distinct from one another. For this reason, we recommend reducing our 3 × 3 matrix from nine scenarios to five. We anticipate that, in most cases, the additional intermediate values created in a 3 × 3 matrix (the grey squares in Fig. 4.5) will not be sufficiently differentiated from the extreme values (Scenarios 1–4) to warrant their own separate scenario narratives. However, we suggest that there is some benefit in augmenting the IL2 × 2 structure with one additional scenario that describes a medium of both critical uncertainties (Scenario 5). We shall call this augmented model 'IL 5'.

The advantage of IL 5 is that, as well as bounding the range of plausible outcomes at the extremes of each variable, we are also showing what a medium value might look like, without creating an 'illusion of certainty' (O'Mahony 2014, p. 42). This is part of the broad-brush prediction service that we recommend scenarios provide. A benefit is that we retain the usual set of well-contrasted storylines (Scenarios 1–4), which force participants outside their usual perspectives and assumptions. At the same time, we can develop a narrative that sits in the middle of the probability spectrum (Scenario 5), but it must be used with discretion. It should not overshadow the rest and give false confidence in a likely path.

4.20 Summary

From our review of scenario *objectives*, *approaches,* and *techniques*, we suggest that scenario methods can be grouped under one or more of seven broad approaches: *deductive*, *inductive*, *quantitative*, *hybrid*, *integrated*, *scoping,* and *engagement*. Each has its strengths and (at times, considerable) weaknesses. We conclude that no single scenario approach can match the scale and complexity of Antarctic geopolitics. For that reason, we recommend that a scenario process employ a judicious blend of several approaches. This is more likely to deliver scenarios that are plausible, rigorous, and robust.

We then narrowed our gaze to a single technique from the intuitive logics (IL) process: the *standard IL 2 × 2 matrix* commonly deployed in geopolitical scenarios. To test some of the strengths and weaknesses of this technique, we unpacked the process over nine steps, using a topical example from Antarctic geopolitics—resource extraction. In doing so, we encountered some difficulties with this technique (at least in its standard formulation) which may be relevant for scenario practice in general. This has led us to propose some modifications to the scenario planning process, to provide a more nuanced analysis of driving forces. Drawing on our earlier argument that part of the services that scenarios provide is targeted, broad-brush prediction, we propose the introduction of a fifth scenario (the 'IL 5' model), in which, as well as bounding the range of plausible outcomes at the high and low extremes of each variable (with 2 × 2), we also show what a scenario based on medium values of the two driving forces might look like. In doing so, we are careful not to imply that this fifth scenario is more probable than the rest.

Scenario development provides a framework to structure our thinking on the future. Avoiding ad-hoc, arbitrary, implausible, or illogical scenarios requires careful consideration of the development and interaction of driving forces. The scenario axes these drivers create should be viewed with caution as a structural tool, and not as a rigid blueprint or 'straitjacket' (O'Mahony 2014, p. 48); they are ways to generate rich, multidimensional conceptions of the future. For us, a key point is that ostensibly practical decisions about scoping and framing of scenarios are in fact shaped—in important ways—by the assumptions and values we hold. Indeed, scenario processes are not the objective, technical considerations they often are perceived to be. Finally, whatever our objectives, approaches and techniques, a successful scenario process will include an engagement strategy which is highly iterative; which communicates effectively; and which is practically useful to the sponsoring organisation and intended audiences.

References

Amer, Muhammed, Tugrul U. Daim, and Antonie Jetter. 2013. A Review of Scenario Planning. *Futures* 46: 23–40.

Baena, Alethea B. M. 2017. Gaming Futures: To Experience Scenarios Through *Teatro Del Devenir*. *Journal of Futures Studies* 22 (2): 119–124.

Barker, Alec D. 2014. Applying Structured Scenario Fusion to Address Multifaceted Strategic Challenges. *Journal of the Operational Research Society* 65 (11): 1726–1734.

Bell, Benjamin, and Michael Marks. 2016. Composite Signatures Analyst Learning Tool (CSALT): Supporting the Analyst with Scenario-Based Methodology Training. *The International Journal of Intelligence, Security, and Public Affairs* 18 (2): 157–172.

Berlin, Gordon, and Robert Solow. 2009. Get the question right, then choose the method. *Journal of Policy Analysis and Management* 28: 175–6.

Bezold, Clement. 2009. Aspirational Futures. *Journal of Futures Studies* 13 (4): 81–90.

Billen, Gilles, Julia Le Noë, and Josette Garnier. 2018. Two Contrasted Future Scenarios for the French Agro-Food System. *Science of the Total Environment* 637–638: 695–705.

Bradfield, Ron, George Wright, George Burt, George Cairns, and Kees Van der Heijden. 2005. The Origins and Evolution of Scenario Techniques in Long Range Business Planning. *Futures* 37 (8): 795–812.

Bradfield, Ronald, James Derbyshire, and George Wright. 2016. The Critical Role of History in Scenario Thinking: Augmenting Causal Analysis Within the Intuitive Logics Scenario Development Methodology. *Futures* 77: 56–66.

Cairns, George, Iftekhar Ahmed, Jane Mullett, and George Wright. 2013. Scenario Method and Stakeholder Engagement: Critical Reflections on a Climate Change Scenarios Case Study. *Technological Forecasting and Social Change* 80 (1): 1–10.

Cairns, George, George Wright, Peter Fairbrother, and Richard Phillips. 2017. 'Branching Scenarios' Seeking Articulated Action for Regional Regeneration—A Case Study of Limited Success. *Technological Forecasting and Social Change* 124: 189–202.

Chermack, Thomas J. 2018. An Analysis and Categorization of Scenario Planning Scholarship from 1995–2016. *Journal of Futures Studies* 22 (4): 45–59.

Chermack, Thomas J., and Laura Coons. 2015. Scenario Planning: Pierre Wack's Hidden Messages. *Futures* 73: 187–193.

Cohen, Michael. 1995. Induction. In *The Oxford Companion to Philosophy*, ed. Ted Honderich, 405–406. Oxford: Oxford University Press.

Cresser, J. D. 2011. *Quantum Physics Notes*. Sydney: Macquarie University. https://physics.mq.edu.au/~jcresser/Phys304/Handouts/QuantumPhysicsNotes.pdf. Accessed July 22, 2020.

Dator, Jim. 2009. Alternative Futures at the Manoa School. *Journal of Futures Studies* 14 (2): 1–18.

Derbyshire, James, and George Wright. 2017. Augmenting the Intuitive Logics Scenario Planning Method for a More Comprehensive Analysis of Causation. *International Journal of Forecasting* 33 (1): 254–266.

Durance, Phillipe, and Michel Godet. 2010. Scenario Building: Uses and Abuses. *Technological Forecasting and Social Change* 77 (9): 1488–1492.

Fink, Alexander, Andreas Siebe, and Jens-Peter. Kuhle. 2010. How Scenarios Interconnect Strategy, Innovation, and Early Warning Processes. *World Futures Review* 2 (1): 5–30.

Freeth, Rebecca, and Scott Drimie. 2016. Participatory scenario planning: from scenario 'Stakeholders' to scenario 'Owners'. *Environment: Science and Policy for Sustainable Development* 58(4): 32–43.

Gambelli, Danilo, Daniela Vairo, and Rafaelle Zanoli. 2010. Exploiting Qualitative Information for Decision Support in Scenario Analysis. *Journal of Decision Systems* 19 (4): 407–422.

Glenn, Jerome C., Theodore J. Gordon, and Elizabeth Florescu. 2010. *2010 State of the Future*. Washington, D. C.: The Millennium Project.

Godet, Michel. 1986. Introduction to *La Prospective*: Seven Key Ideas and One Scenario Method. *Futures* 18 (2): 134–157.

Godet, Michel, and Fabrice Roubelat. 1996. Creating the Future: The Use and Misuse of Scenarios. *Long Range Planning* 29 (2): 164–171.

Green, Kesten C., and J. Scott Armstrong. 2011. Role-thinking: Standing in other people's shoes to forecast decisions in conflicts. *International Journal of Forecasting* 27(1): 69–80.

Halim, Ronald A., Jan H. Kwakkel, and Lóránt. A. Tavasszy. 2016. A Scenario Discovery Study of the Impact of Uncertainties in the Global Container Transport System on European Ports. *Futures* 81: 148–160.

Hasegawa, Yasuhiro, Kimiko Okabe, and Hisatomo Taki. 2018. A Scenario Approach for Ecosystem-Service Changes. *Futures* 96: 23–31.

Hemmings, Alan D. 2013. Utopian Framings of the Polar Regions. *The Polar Journal* 3 (2): 273–276.

Huss, William R. 1988. A Move Toward Scenario Analysis. *International Journal of Forecasting* 4 (3): 377–388.

Kahane, Adam. 2012. *Transformative Scenario Planning: Working Together to Change the Future*. Oakland, CA: Berrett-Koehler Publishers Inc.

Kahn, Herman, and Anthony J. Wiener. 1967. *The Year 2000: A Framework for Speculation on the Next Thirty-Three Years*. New York: MacMillan.

Kaufmann, Petra R. 2016. Integrating Factor Analysis and the Delphi Method in Scenario Development: A Case Study of Dalmatia, Croatia. *Applied Geography* 71: 56–68.

Kharrazi, Ali, and Masahiro Kakuwa. 2017. Scenario Projects in Japanese Government: Strategic Approaches for overcoming Psychological and Institutional Barriers. *Futures* 86: 18–26.

Kok, Kasper. 2016. *Multi-Scale Integration and Synthesis of Scenarios and Adaptation Narratives*. ECONADAPT Deliverable 1.5. Wageningen: Wageningen University. https://econadapt.eu/sites/default/files/docs/Deliverable%201-5%20approved%20for%20publishing.pdf. Accessed July 20, 2021.

Kok, Kasper, and Hedwig van Delden. 2009. Combining two approaches of integrated scenario development to combat desertification in the Guadalentin watershed, Spain. *Environment and Planning B: Planning and Design* 36: 49–66.

Konno, Noboru, Ikujiro Nonaka, and Jay Ogilvy. 2014. Scenario Planning: The Basics. *World Futures: The Journal of General Evolution* 70 (1): 28–43.

Lacey, Alan R. 1995. Intuition. In *The Oxford Companion to Philosophy*, ed. Ted Honderich, 415. Oxford: Oxford University Press.

Leane, Elizabeth. 2013. Yesterday's Tomorrows and Tomorrow's Yesterdays: Utopian Literary Visions of Antarctic Futures. *The Polar Journal* 3 (2): 333–347.

Leigh, Andrew. 2003. Thinking Ahead: Strategic Foresight and Government. *Australian Journal of Public Administration* 62 (2): 3–10.

Leseure, Michel. 2019. Is the Recovery of Failing Scenario Analysis a Legitimate and Valuable Activity? *World Futures Review* 11 (1): 80–97.

Liggett, Daniela, Bob Frame, Neil Gilbert, and Fraser Morgan. 2017. Is it All Going South? Four Future Scenarios for Antarctica. *Polar Record* 53 (5): 459–478.

Lindgren, Mats, and Hans Bandhold. 2009. *Scenario Planning: The Link Between Future and Strategy*, 2nd ed. Basingstoke, UK: Palgrave Macmillan.

Lord, Steven, Ariella Helfgott, and Joost M. Vervoort. 2016. Choosing Diverse Sets of Plausible Scenarios in Multidimensional Exploratory Futures Techniques. *Futures* 77: 11–27.

Mack, Ted C. 2010. Memes in Foresight: Response to the Black Swan. *World Futures Review* 2 (2): 5–15.

Marcus, Ruth B. 1995. Deduction. In *The Oxford Companion to Philosophy*, ed. Ted Honderich, 181. Oxford: Oxford University Press.

Martelli, Antonio. 2001. Scenario Building and Scenario Planning: State of the Art and Prospects of Evolution. *Futures Research Quarterly* 17 (2): 57–70.

Martin, Michael G. F. 1995. Imagination. In *The Oxford Companion to Philosophy*, ed. Ted Honderich, 395. Oxford: Oxford University Press.

Mearns, L.O., M. Hulme, T.R. Carter, R. Leemans, M. Lal, P. Whetton, L. Hay, et al. 2001. Climate Scenario Development: Contribution Of Working Group I to the Third Assessment Report of the Intergovernmental Panel on Climate Change. In *Climate Change 2001: The Scientific Basis*, ed. J.E.T. Houghton, Y. Ding, D. Griggs, M. Noguer, P. Van der Linden, X. Dai, M. Maskell, et al., 739–768. Cambridge: Cambridge University Press.

Meissner, Philip, Christian Brands, and Torsten Wulf. 2017. Quantifying Blind Spots and Weak Signals in Executive Judgment: A Structured Integration of Expert Judgment into the Scenario Development Process. *International Journal of Forecasting* 33 (1): 244–253.

Mercer, David. 1995. Scenarios Made Easy. *Long Range Planning* 28 (4): 7–86.

Miller, Mel, and Ian Siggins. 2003. A Framework for Intergenerational Planning. *Foresight* 5 (6): 18–25.

Nesbit, R., S. Moulton, S. Robinson, C. Smith, L. DeHart-Davis, M.K. Feeney, B. Gazley, and Y. Hou. 2011. Wrestling with Intellectual Diversity in Public Administration: Avoiding Disconnectedness and Fragmentation While Seeking Rigor, Depth, and Relevance. *Journal of Public Administration Research and Theory* 21 (suppl_1): i13–i28.

Nicosia, Manuela. 2011. Scenario-Based Strategy: An Interview with Tim Mack, President of the World Future Society. *World Futures Review* 3 (4): 63–70.

Nygrén, Nina A. 2019. Scenario Workshops as a Tool for Participatory Planning in a Case of Lake Management. *Futures* 107: 29–44.

O'Mahony, Tadhg. 2014. Integrated Scenarios for Energy: A Methodology for the Short Term. *Futures* 55: 41–57.

Oxford English Dictionary [Online]. 2021. Oxford: Oxford University Press. https://www.oed.com. Accessed July 15, 2021.

Paas, Fred, and Jeroen J. G. van Merriënboer. 2020. Cognitive-Load Theory: Methods to Manage Working Memory Load in the Learning of Complex Tasks. *Current Directions in Psychological Science* 29 (4): 394–398.

Paya, Ali. 2018. Critical Rationalism as a Theoretical Framework for Futures Studies and Foresight. *Futures* 96: 104–114.

Ramírez, Rafael, and John W. Selsky. 2016. Strategic Planning in Turbulent Environments: A Social Ecology Approach to Scenarios. *Long Range Planning* 49 (1): 90–102.

Ramírez, Rafael, and Angela Wilkinson. 2014. Rethinking the 2×2 Scenario Method: Grid or Frames? *Technological Forecasting and Social Change* 86: 254–264.

Ramírez, Rafael, and Angela Wilkinson. 2016. *Strategic Reframing: The Oxford Scenario Planning Approach*. Oxford: Oxford University Press.
Ramírez, Rafael, Malobi Mukherjee, Simona Vezzoli, and Arnoldo Matus Kramer. 2015. Scenarios as a Scholarly Methodology to Produce 'Interesting Research.' *Futures* 71: 70–87.
Riddell, Graeme A., Hedwig Van Delden, Graeme C. Dandy, Aaron C. Zecchin, and Holger R. Maier. 2018. Enhancing the Policy Relevance of Exploratory Scenarios: Generic Approach and Application to Disaster Risk Reduction. *Futures* 99: 1–15.
Ritchey, Tom. 2011. Modeling Alternative Futures with General Morphological Analysis. *World Futures Review* 3 (1): 83–94.
Schnaars, Steven P. 1987. How to Develop and Use Scenarios. *Long Range Planning* 20 (1): 105–114.
Schoemaker, Paul J. H. 1993. Multiple Scenario Development: Its Conceptual and Behavioral Foundation. *Strategic Management Journal* 14 (3): 193–213.
Schwartz, Peter. 1996. *The Art of the Long View: Planning for the Future in an Uncertain World*. Chichester, UK: Wiley & Sons.
Sharma, Ravi S., and Yi Yang. 2015. A Hybrid Scenario Planning Methodology for Interactive Digital Media. *Long Range Planning* 48(6): 412–429.
Spaniol, Matthew J., and Nicholas J. Rowland. 2018. The Scenario Planning Paradox. *Futures* 95: 33–43.
Talberg, Anita, Sebastian Thomas, and John Wiseman. 2018. A Scenario Process to Inform Australian Geoengineering Policy. *Futures* 101: 67–79.
Taleb, Nassim N. 2010. *The Black Swan: The Impact of the Highly Improbable*. New York: Random House.
Tapinos, Efstathios. 2012. Perceived Environmental Uncertainty in Scenario Planning. *Futures* 44: 338–345.
Tapio, Petri, Heidi Rintamäki, Pasi Rikkonen, and Juho Ruotsalainen. 2017. Pump, Boiler, Cell or Turbine? Six Mixed Scenarios of Energy Futures in Farms. *Futures* 88: 30–42.
Totin, Edmond, James R. Butler, Amadou Sidibé, Samuel Partey, Philp K. Thornton, and Ramadjita Tabo. 2018. Can Scenario Planning Catalyse Transformational Change? Evaluating a Climate Change Policy Case Study in Mali. *Futures* 96: 44–56.
United Nations Industrial Development Organization. 2005. *UNIDO Technology Foresight Manual (Vol. 1)*. Vienna: UNIDO. https://open.unido.org/api/documents/4704632/download/UNIDO-Publication-2005-4704632. Accessed August , 2021.
Van der Heijden, Kees. 1996. *Scenarios: The Art of Strategic Conversation*. Chichester, UK: Wiley & Sons.
Varho, Vilja, and Petri Tapio. 2013. Combining the Qualitative and Quantitative with the Q2 Scenario Technique—The Case of Transport and Climate. *Technological Forecasting and Social Change* 80 (4): 611–630.
Wack, Pierre. 1985a. Scenarios: Shooting the Rapids. *Harvard Business Review* (November–December 1985): 139–150.
Wack, Pierre. 1985b. Scenarios: Uncharted Waters Ahead. *Harvard Business Review* (September–October 1985): 73–89.
Wodak, Josh, and Timothy Neale. 2015. A Critical Review of the Application of Environmental Scenario Exercises. *Futures* 73: 176–186.
Wright, George, Ron Bradfield, and George Cairns. 2013. Does the Intuitive Logics Method—and its Recent Enhancements—Produce 'Effective' Scenarios? *Technological Forecasting and Social Change* 80 (4): 631–642.
Young, Oran R. 2002. *The Institutional Dimensions of Environmental Change: Fit, Interplay and Scale*. Cambridge, MA: The MIT Press.
Young, Oran R. 2010. *Institutional Dynamics: Emergent Patterns in International Environmental Governance*. Cambridge, MA: MIT Press.
Zentner, René. 1982. Scenarios, Past, Present and Future. *Long Range Planning* 15 (3): 12–20.

Chapter 5
Antarctic Geopolitics: Background

Abstract Antarctica and the Southern Ocean are largely absent from the leading early works of classical geopolitics. While more recent works have embraced the southern hemisphere, their gaze rarely ventures toward the frozen south. We believe the time is ripe for classical analysis to be re-enlivened in Antarctic geopolitics, and in scenario analysis. In this chapter, we survey some of the key driving forces of Antarctic geopolitics, through a classical lens and across nine themes. From this snapshot, we observe that geopolitical driving forces tend to differ according to whether they are predetermined elements or critical uncertainties; whether they impel events or trends forward or constrain them; whether they instigate rivalry or promote cooperation; and whether they are endogenous or exogenous to the field in question. We suggest that unpacking our driving forces in this way provides us with a more nuanced understanding of how these factors are likely to behave within a scenario.

Keywords Antarctic geopolitics · Disciplines of geopolitics · Driving forces of geopolitics · Driving forces of Antarctic geopolitics · Dynamic signature of driving forces · Scenario planning challenges

5.1 Overview

In this chapter, we shift our focus away from scenario approaches and toward the factors likely to drive and shape future state activity in Antarctica. As we saw in Chap. 4, these driving forces are the core elements that move scenario narratives forward, and influence how we formulate and handle geopolitical issues within them. If we are to understand how driving forces work within a scenario narrative, we need to grasp their origins and dynamics. In this chapter, we will consider both. If our scenarios are to be plausible, rigorous, and robust, then our list of driving forces needs to be comprehensive and prioritised strategically, with key causal relationships between them identified. To be comprehensive, we propose that it helps to look at our subject matter through the lenses of nine themes: Politics, Economics, Social, Technological, Environmental and Legal (the PESTEL categories) (Wright et al. 2013; Meissner et al. 2017; Derbyshire and Wright 2017; Cairns et al. 2013; Bradfield et al. 2016; Augier et al. 2018; Agustan et al. 2019) *as well as* Sloan's three core

J. McGee et al., *The Future of Antarctica*, Springer Polar Sciences,
https://doi.org/10.1007/978-981-16-7095-4_5

disciplines of geopolitics, which we encountered in Chap. 1: geography, strategic studies and history.

5.2 A Brief Recap: Geopolitics—Classical, Critical, and Other Approaches

In Chap. 1, we noted that geopolitics concerns the relationship between the geographic setting to international political power (Van Efferink 2017). In this book, our particular focus is on the intersection of power, influence, and space in the rivalry between states (Powell and Dodds 2014; Sloan 2017). Following Sloan, we observed the multi-disciplinary roots of geopolitics in geography, strategic studies, and history, and suggested that the term 'spatial rivalry' perhaps captures the context of Antarctic geopolitics better than notions of 'conflict'. In Antarctica, the risk of overt conflict has been reduced greatly by 60 years of non-militarisation and the bespoke arrangements to handle claims for territorial sovereignty. This is unpacked further in Chap. 6. In Chap. 1, we noted the contrasts between the two major schools of geopolitics. We saw that *classical* geopolitics (actually a grouping of theories (Flint and Mamadouh 2015)) is grounded in an epistemic assumption about the objectivity of phenomena (Sloan 2017) and takes a primarily state-centric perspective (Powell and Dodds 2014) about control over geographical territory (Wehrmann 2019). On the other hand, *critical* geopolitics (again an aggregation of perspectives) emphasises the subjective nature of geopolitical knowledge, and often elevates the voices of non-state actors (Powell and Dodds 2014). It critiques or deconstructs geopolitical artefacts to uncover the assemblages of power and knowledge used to assert control over spaces (Sloan 2017; Wehrmann 2019). These artefacts include foreign policy documents, speeches, symbols and the language of justification (Orakhelashvili 2008), as well as practices, materials and emotions (Jones and Clark 2019). Critical geopolitics takes a normative stance of bringing to light these assemblages of power and these framings of knowledge, and (in particular) through enabling the voices of marginalised interests to be heard.

Also in Chap. 1, we affirmed Kelly's definition of classical geopolitics as 'the study of the impact or influence of certain geographic features—these being positions and locations of regions, states, and resources plus topography, climate, distance, immigration, state's sizes and shapes, demography and the like—upon states foreign policies and actions as an aid to statecraft' (Kelly 2016, p. ix). We suggest that much of this formulation applies to the circumstances of Antarctica and is broadly consistent with Sloan's typology for the three core disciplines of geopolitics, noted above. We also questioned whether geopolitical analysis must be limited to a binary distinction between classical and critical approaches and suggested, instead, that *critical realism* offers a prospective 'third way' to harness the benefits of both schools for scenario development. We do not develop this further here; however, we do suggest that critical realism is worth serious investigation as a pragmatic and potentially valuable approach to the theory and practice of geopolitical scenarios.

We may be swimming against the academic tide of the last few decades, but in this book we have chosen a classical geopolitical approach to explore scenario planning. We believe there are good reasons to do so. In the absence of a permanent human population, the role of the state is much stronger in Antarctica than in most other regions of the planet. The central human activity and majority of financial investment on the Antarctic continent is to undertake and support scientific research. Due to the severe climatic and logistical difficulties of the region, most of that science is carried out by national Antarctic programs. While significant activity is conducted by non-state actors in tourism and fisheries, the central actors in governance of Antarctica are states, acting through the regular meetings and diplomatic practices of the Antarctic Treaty System (ATS). As critical geopolitical analysis of Antarctica has flourished over recent decades, we believe a gap is developing in academic analysis using the classical approach. Further, as the world *above* 60 degrees South moves into a new period of great power competition, the perceived role and importance of the state is strengthening in international affairs. In our view, it is timely to reapply the classical geopolitical lens to Antarctica.

5.3 Antarctica and Classical Geopolitical Analysis

Unsurprisingly, Antarctica and the Southern Ocean are largely absent from the leading early works of classical geopolitics. Mackinder's seminal contribution, *The Geographical Pivot of History* (Mackinder 1904), makes no mention of Antarctica. Instead, Mackinder's analysis centres on the northern hemisphere 'pivot area' of the Eurasian continent, as well as the partly continental/partly oceanic 'inner crescent' zone surrounding it. Mahan [1890 (2010a) and 1892 (2010b)] does consider a wholly oceanic 'outer crescent' zone, extending south to the tip of Africa and bottom of Australia—but no further. In his view, however, this was the least important zone for world politics. In similar vein, Spykman's classic *Geography of the Peace* (Spykman and Nicholl 1944) contains brief commentary and a map on polar air routes, but these relate only to the Arctic. Martin's 1949 chapter, 'The Antarctic Sphere of Interest', examines the political, economic and strategic geographies of post-war Antarctica, prior to the Antarctic Treaty. At the political level, he details the bases of a future US claim to 'American Antarctica' (Martin 1949, pp. 66–73). These include acts of discovery, and application of the 1823 Monroe Doctrine to the portion of the continent that lies within the western hemisphere. The economic potential of Antarctica's oil, coal, and other minerals is vast, but cannot be realised in the near term. In the event of 'another war', Martin concludes that Antarctica is 'of relatively little strategic importance' (Martin 1949, p. 77).

While more recent general works on classical geopolitics have embraced issues in the southern hemisphere, their gaze rarely ventures toward the frozen south. For instance, Cohen's 2014 edition of *Geopolitics: The Geography of International Relations* (Cohen 2014) contains chapters on South America and the Asia–Pacific rim but, aside from a brief mention of the Falklands War, there is no substantial analysis of Antarctica or the Southern Ocean. In contrast, there are multiple references to issues in the Arctic. Rare exceptions to this pattern are the classically oriented contributions of Child in *Antarctica and South American Geopolitics: Frozen Lebensraum* (Child 1988) and the edited volume by Kelly and Child, *Geopolitics of the Southern Cone and Antarctica* (also 1988) (Kelly and Child 1988). Written during the intense period when the proposed minerals regime for Antarctica was well advanced but not yet abandoned, these works survey the geopolitical interests of various South American states in Antarctica. Arguably, Brady continues to carry the classical torch (Brady 2013a, 2013b, 2017a, 2017b), most notably with her analyses of Chinese interests and actions within the Antarctic Treaty Area (ATA). We believe the time is ripe for classical analysis to be re-enlivened in Antarctic geopolitics, and particularly in scenario analysis.

5.4 Nine Themes for the Driving Forces of Antarctic Geopolitics

As we saw in Chaps. 3 and 4, a key step in creating scenarios is to identify the major *driving forces*. In scenarios using deductive reasoning, these are the premisses from which scenario conclusions follow. They are causal factors important to understanding the field in question, as well as being 'the elements that move the plot of a scenario' forward (Schwartz 1996, pp. 101–102). So, how do we ensure that we capture a sufficient breadth of driving forces at play within a given field, so that our scenarios are not unduly narrow in their focus? To ensure that most relevant macro-scale dimensions of the external environment are considered, driving forces are often grouped within six 'PESTEL' themes: Politics, Economics, Social, Technological, Environmental and Legal) (Wright et al. 2013; Meissner et al. 2017; Derbyshire and Wright 2017; Cairns et al. 2013; Bradfield et al. 2016; Augier et al. 2018; Agustan et al. 2019).[1] For scenarios with a specific focus on geopolitics, van Efferink proposes that the traditional PESTEL categories be supplemented with the facts and perceptions of *geography* (both physical and human).[2] We concur—particularly given the exceptional impacts of Antarctica's physical geography on its geopolitics.

Building on this, we argue there is value in adding Sloan's two other disciplines of geopolitics as specific areas of focus when identifying driving forces: *strategy* and *history*. For Sloan, in the context of geopolitics, *strategy* is 'not merely a static plan of action, rather it is a dynamic awareness of one's goals, an evaluation of the efficacy of means, and an appreciation of consequences' (Sloan 2017, p. 36). Strategy is 'neither

[1] Also referred to as PESTLE (Liggett et al. 2017), with other variations.

[2] "Geopolitical Scenario Planning", Course Notes, Maastricht University Summer School 2019, unpublished.

policy nor armed combat; rather it is the bridge between them' (Sloan 2017, p. 36). A focus on strategy adds depth to our analysis of state interests and goals, the extent to which these interests coalesce or diverge, and how these dynamics affect states' commitment to the rules, norms and values of the international system.

For Sloan, *history* is 'an appreciation and utilization of temporal frameworks that enable an observer to describe and analyse conflicts that have occurred in the past or break out in the present, and possibly even predict their future emergence and resolution' (Sloan 2017, p. 37). So, why should we bother with *historical* factors as a specific theme of Antarctic geopolitics? Bradfield and colleagues warn of the dangers we face if we overlook the ongoing influences (and learnings) from the past:

> Templates such as PESTEL focus on what is salient at present, providing little room for consideration of how things have developed over time up to the present, how one identified driving force may have gained in prominence over time, or how past significant changes or surprises were the result of particular important driving forces. It does not take account of how causal factors were wrongly interpreted by contemporaries at the time at which surprise outcomes occurred in the past. A great deal of potential learning from how previous expectations were misguided is overlooked. It stands to reason, then, that scenario planning conducted in this way does little to assist in overcoming the repeated failure to learn from the past. (Bradfield et al. 2016, p. 62)

Indeed, we suggest that a dedicated focus on the historical antecedents of current geopolitical factors is indispensable to our analysis. This extends to historical 'facts' and perceptions. Why consider these? As we saw in Chap. 3, there are definite limits on our capacity to predict the future, given the combined operation of complexity, uncertainty and time. However, this *does not* imply a complete loss of causal connection between the past, present and future. Even in times of turbulence, novelty and ambiguity (Ramírez and Wilkinson 2016), some factors from the past are likely to have ongoing causal influence on the path ahead. As noted in Chap. 2, for example, the exploits of the 'Heroic Era' continue to exert an influence in the geopolitics of the frozen continent.

For all these reasons, we propose a total of nine themes or focus areas for analysis of the driving forces of geopolitics—see Fig. 5.1. These themes are ways to organise information and apply disciplinary lenses to it; they are not discrete categories.

5.5 Challenges for the Scenario Planner!

Undertaking this analysis poses various challenges for the scenario planner. The first challenge is that each of these themes will have its own language, epistemology, approaches, and outputs (Proctor and Vu 2019). Integrating these into a coherent, internally consistent scenario narrative can be difficult. As Proctor and Vu observe, however, there are often sufficient methods in common between neighbouring disciplines to enable fruitful collaboration (Proctor and Vu 2019). A second practical challenge is to identify and harness sufficient expertise across these diverse disciplines. The composition of the core team and external experts is pivotal here. A third

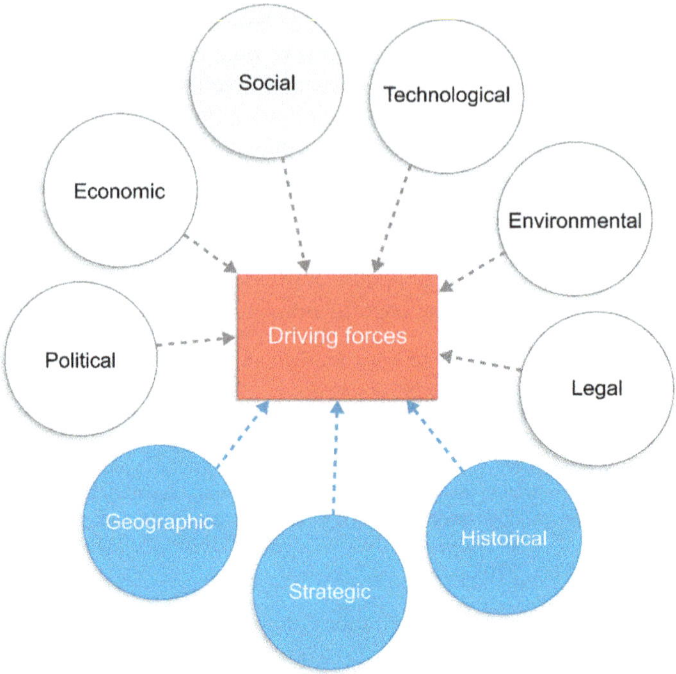

Fig. 5.1 Nine themes for driving forces: PESTEL + three dsiciplines of geopolitics

challenge is the risk that, if we pre-judge which of these themes we will pay most attention to, our scenarios may end up reinforcing our own biases and overlooking key driving forces at work in a given field. A conscious striving for a balanced focus across these themes could reduce this risk. A fourth challenge is that, if our scenarios are to achieve robustness and rigour, this analysis cannot be done in five minutes. The sponsor needs to invest sufficient time and resources so that relevant data can be sourced and analysed properly. To illustrate, Derbyshire and Wright note that a typical scenario exercise may identify over 200 distinct driving forces across the PESTEL themes (Derbyshire and Wright 2017). Finally, at least some of these themes are likely to impact each other. Accordingly, some time and effort should also be allocated to identifying key causal links *between* these themes (Wright et al. 2013). None of these challenges is insurmountable. However, if the preparatory analysis is narrow, superficial and rushed, the ensuing scenarios are unlikely to be of lasting value. Before turning to the nine themes of Antarctic geopolitics, we will address some general dynamics of the driving forces that we have encountered.

5.6 Four Dynamics of Driving Forces

Our survey of the driving forces of Antarctic geopolitics reveals some *dynamics of driving forces* which may be relevant not just for Antarctic scenarios, but for scenarios in general. It became clear that not all driving forces operate in the same way within our scenarios. They can vary profoundly in terms of the extent of our knowledge about them and in their causal impacts. Specifically, we found that driving forces of Antarctic geopolitics tend to differ according to: whether they are *critical uncertainties* or *predetermined elements*; whether they *impel* events or trends forward, or *constrain* them; whether they are *exogenous* or *endogenous* to the field in question; and whether they instigate *rivalry*, or promote *cooperation.*

We will describe each of these dynamics below. We suggest that unpacking our driving forces in this way provides a more nuanced understanding of the causal impacts they are likely to have within a scenario. This can be useful in two ways. First, it may help clarify (and, if necessary, challenge) the assumptions that scenario participants may hold about the impact of a particular driving force. Second, it can help ensure that the decisions we make about the trajectory of a scenario narrative are internally consistent with our assumptions. We will revisit dynamic signatures in our case study on militarisation, in Chap. 7.

Three technical points arise as we consider the dynamics of driving forces. The first is that not every stakeholder will assess these dynamics in the same way. So, for internal consistency, we suggest that the dynamics of driving forces be assessed from the perspective of our chosen 'scenario standpoint' (see Chap. 4). As we saw, this is the viewpoint of the actor (or group of actors) whose interests, strategies and tactical decisions will 'stress tested' as our scenarios unfold. The second point is that, if using the IL 2 × 2 method (Chap. 4), it is worth checking whether these dynamics differ when the driving force is instantiated at its **high** or **low** level. The third point is that our initial assessments of the dynamics of a driving force may need to change as the scenario evolves: perhaps a predetermined element becomes more uncertain; perhaps a driver for cooperation ends up as a driver for rivalry. For this reason, we recommend starting with some evidence-based assumptions, then periodically re-assessing whether these dynamics still hold as the scenario progresses.

1. **Driving forces as *predetermined elements or critical uncertainties***
   The first dynamic of driving forces is an elaboration of the distinction (noted in Chap. 4) between predetermined elements and critical uncertainties. As we saw, *predetermined elements* (Schwartz 1996; Amer et al. 2013) are significant factors which are already known; easy to predict; or assumed to be relatively stable into the future. In contrast, we saw that *critical uncertainties* (Schwartz 1996; Van der Heijden 1996) are factors which are highly unpredictable; unmeasurable; likely to have a pivotal impact; and difficult to influence. We noted that boundaries between predetermined elements and critical uncertainties can be hard to draw and can change over time. For this reason, we suggested that, rather than categorising driving forces as critical uncertainties *or* predetermined elements, they could be considered somewhere along a spectrum.

2. **Driving forces as *impelling* or *constraining***
The second dynamic of driving forces is the distinction between those driving forces that drive trends and events forward (*impelling*), and those that hold them back *(constraining*). In her work on contemporary security concerns in the Antarctic, Nyman refers to ways in which security on the continent is considered to be *achieved* or *compromised* (Nyman 2017, p. 571).This suggests that some driving forces work in opposite directions in terms of achieving Antarctic security. There is another dynamic at work, however, which is less obvious. From the literature on the geopolitics of Antarctic minerals, it seems that quite a number of driving forces are not, of themselves, 'driving' in the sense of '[i]mpelling, setting in motion, actuating' (Oxford English Dictionary [Online] 2021). These are factors which have no impelling force of their own, but which can play a vital role in the outcome of events. We refer to this type of driving force as a *constraint* (Sloan 2017; Hemmings 2012). As noted above, in some cases a driving force may be impelling for one party and constraining for another—here, we assess this from the perspective of our 'scenario standpoint'.
Why is this distinction important? Scanning for both types of driving force is useful for scenario practice because constraints may exhibit quite different dynamics from impelling driving forces: on average, they may be more predictable, less turbulent—and less obvious. If we only scan for *impelling* driving forces, we risk overlooking a whole class of causal dynamics (i.e., *constraints*) that operate quietly, actively, and, at times, significantly, in the background. To illustrate, Antarctic ice sheets impel nothing in a geopolitical sense; they are inert. However, by covering 98 per cent of the continent at an average thickness of 2.5 km, ice sheets operate as a formidable practical constraint on driving forces for Antarctic mining (Orakhelashvili 2008; Abdel-Motaal 2016). Likewise, *if* there were large-scale reductions in Antarctic ice sheets due to global warming (Steig 2019), this would reduce the ice's role as a constraint on future mineral extraction. We suggest this distinction between impelling driving forces and constraints benefits geopolitical scenarios more broadly. While the causal impacts of constraints (particularly geographic constraints) are recognised within the specific literature on Antarctica (Orakhelashvili 2008; Dodds and Nutall 2016), this dynamic appears less frequently in the general literature on scenarios. This has led us to question whether scenario practice in other fields is scanning sufficiently for this important, but underrated, dynamic.
3. **Driving forces as *endogenous* or *exogenous***
From the literature, it is also clear that some driving forces and constraints are *endogenous* to Antarctica and the ATS, and some are *exogenous.* With this distinction in mind, Antarctic geopolitical scenarios can draw on models such as Young's institutional dynamics (Young 2010) to probe for alignment and misalignment between the ATS regime's *internal* performance against institutional 'good governance' indicators and driving forces in the *external* environment (Young 2010). In some cases, the driving force may have a blend of endogenous and exogenous influences, so a simple binary choice between the two will not suffice. We address this point below.

4. **Driving forces for *rivalry* or *cooperation***

Finally, as we saw in the Introduction (Chap. 1), van Efferink observes that spatial rivalry is a core dynamic of geopolitics. Building on this, we suggest that each driving force, be it impelling or constraining, operates somewhere on a spectrum between *rivalry* [or competition (Axelrod 1997)] and its opposite—*cooperation.* We suggest that making this distinction explicit in geopolitical scenarios is particularly important. Why? Because rivalrous driving forces will have profoundly different causal consequences within a scenario from cooperative driving forces. To this rather obvious point, we add a less evident one that emerged from our survey of Antarctic geopolitics. If we combine our awareness of driving forces as *impelling* or *constraining* with our awareness of driving forces for *rivalry* and *cooperation*, then we have the tools to scan actively for less obvious, but nonetheless crucial, scenario dynamics.

One example is where *rivalry* between states is initially exacerbated not by an *impelling force*, but by a *constraint*. In the case of the ATS, examples of regime constraints include:

- the substantial prerequisites needed for a state to qualify as an Antarctic Treaty Consultative Party[3]—a constraint on full participation;
- the requirement for consensus-based decision-making[4]—a constraint on timely resolution of contentious matters; and
- perceptions that the governance culture of the ATS is 'intensely conservative' and resistant to change[5]—a constraint on reform.

Some of these constraints are necessary compromises to enable the ATS to function, while balancing a diverse range of state interests. An unintended consequence, however, is that the legitimacy and effectiveness of the ATS as a governance regime (Stokke and Vidas 1996) has been called into question. Here, we see a *constraining* driving force provoking an *impelling* driving force. The impelling force favours replacing the ATS with an alternate system of governance, with different membership and different rules for decision-making (Hemmings 2017). We suggest such a move would lead to a more *rivalrous* outcome—a collapse of the current balance between those states interested in the preservation of Antarctica for peace and science, and those states interested in greater exploitation of Antarctica's living and non-living resources.

In summary, we suggest there are vital (and not always obvious) dynamics at work in geopolitical scenarios that can be teased out more clearly if we have the tools to discriminate between those driving forces that provoke *rivalry* and those that promote *cooperation.* Further, we suggest this analysis becomes more powerful when it is combined with other dynamics of driving forces, such as those that are *impelling* or *constraining.*

---

[3] Antarctic Treaty, Art IX(2).

[4] Antarctic Treaty, Art IX(4).

[5] Huber 2011, cited in Liggett (2015), p. 68

5.7 Pulling These Together: The 'Dynamic Signature' of a Driving Force

From our survey of Antarctic geopolitics, we conclude that it is useful to represent each of these dynamics as a spectrum, rather than a binary choice between two extremes. We also suggest that each driving force has an individual 'dynamic signature'. This is simply a graphical representation of where a driving force sits on each the four spectra, as illustrated in Fig. 5.2.

Our first example, Driving Force A, might be global demand for rare earth metals, given Antarctica's known deposits (United States Congress, Office of Technology Assessment 1989, US Geological Survey, National Minerals Information Center 2021). The dynamic signature of Driving Force A (signified by the red dots) indicates that it has a high degree of *uncertainty*; it is an *impelling* force, rather than a constraint on action; it tends to instigate *rivalry* between states; and its origins are largely *exogenous* to the field of Antarctic geopolitics. Our second example, Driving Force B, might be the ongoing commitment of states to observe the rules, norm and values of the Convention on the Conservation of Antarctic Marine Living Resources (CAMLR Convention) regime for Antarctic fisheries management and environmental protection. The dynamic signature of Driving Force B (signified by the yellow dots) indicates that it is also somewhat *uncertain*; that it operates primarily as a *constraint* on unauthorised state action; that it has *exogenous* influences (such as global demand for marine protein) as well as *endogenous* aspects (such as CAMLR obligations); and it is intended to promote *cooperation.*

As noted above, we suggest that unpacking our driving forces in this way provides a more nuanced understanding of their likely causal impacts within a scenario.

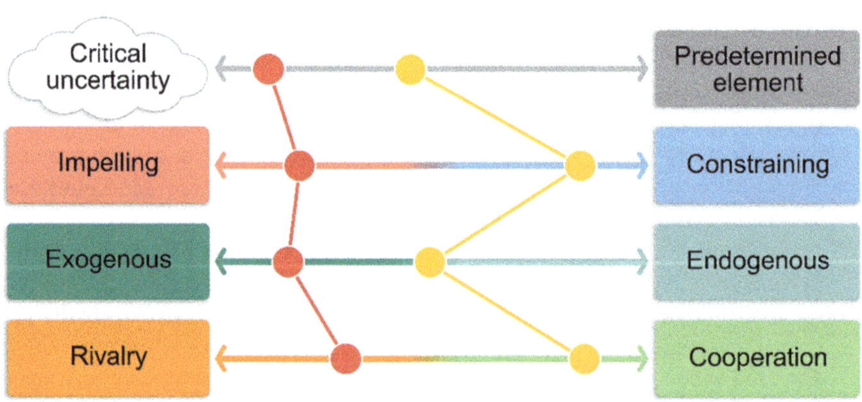

Fig. 5.2 Dynamics of driving forces in Antarctic geopolitics

5.8 Nine Themes for Driving Forces

In this section, we illustrate *some* of the driving forces of Antarctic geopolitics.[6] Here, our goal is not to provide a complete list—this would rapidly become dated. Instead, we consider each theme in turn, and highlight one or two salient topics. Our purpose is to provide a broad contextual overview before narrowing our focus to the driving forces of Antarctic militarisation in Chaps. 6 and 7. Obviously, many issues will span more than one theme, and so their placement within one theme or another is here based on our assessment of 'best fit', rather than being a formal categorisation. As we work through these themes, we will note prominent examples of where driving forces are *endogenous* or *exogenous* to the ATS; where they *impel* or *constrain* the acts of states, their corporations and citizens; where they push states toward *cooperation* or *rivalry*; and where they lie on the spectrum between *critical uncertainty* and *predetermined element*.

1. **Political driving forces**
   The geopolitics of Antarctica are sometimes framed by starkly alternate conceptions: a collection of states claiming sovereign territory (with the exception of the unclaimed Pacific sector) (Collis 2017) *or* as an area beyond national jurisdiction (Hemmings et al. 2017). The Antarctic Treaty provided a settlement of potential political disputes over territory within the ATA through the 'genius' of Article IV. Article IV balances rights and interests between claimants and non-claimant states (noting that the USA and the Soviet Union made reservations over rights to claim 'any or all' of Antarctica). This has allowed a bifocal approach to support ongoing cooperation and collaboration in scientific endeavour. This approach is supported by the Treaty embodying consensus decision-making, which provides stability and legitimacy for agreed outcomes. However, veto power can be wielded by any Antarctic Treaty Consultative Party (ATCP), at times stifling discussion and blocking action on contentious issues (Kaye et al. 2011). Thus, such *endogenous* drivers have broader consequences. As Hemmings and colleagues note, while the ATS is a regionally focused governance regime, the driving forces of Antarctic geopolitics are global in terms of the geographic distribution of actors and consequences (Hemmings et al. 2017). On the other hand, Young suggests that the ongoing resilience of the ATS might be because the regime *has not* changed as much as might be expected (Young 2017).
2. **Economic driving forces**
   Antarctica has been mythologised as an 'El Dorado' of resource riches and economic wealth. This has included arguments over the potential of rare minerals, hydrocarbon reserves (coal, oil, and gas), fisheries and—given concerns over scarcity of potable water at lower latitudes—thoughts of iceberg harvesting. Between 1989 and 1991, the dramatic pivot from the (proposed,

[6] Here, we present these driving forces in their ordinary descriptive form, as factors with potential impact for geopolitical scenarios. They are not expressed in the format necessary in the IL 2 × 2 technique, where a driving force is 'direction-neutral' and then instantiated at high and low levels.

then abandoned) Convention on the Regulation of Antarctic Mineral Resources (CRAMRA) mining regime in favour of the Madrid Protocol's ban on mining (Art 7) cemented the ascendency of a 'preservationist ethic' for Antarctica (Young 2017, p. xv). Fortuitously, the mining ban aligned with the interests of sovereignty claimants who wished to avoid mineral extraction within their territory being perceived as a dilution of their claims (Jackson and Boyce 2011). As Young notes, however, some analysts now believe it is only a matter of time before the 'consumptive juggernaut' overwhelms the preservationist ethic in Antarctica (Young 2017), p. xv]. Closely tied to this are uncertainties about whether (and to what extent) global environmental norms will influence the way in which economic pressures on Antarctica are handled in coming decades (Young 2017).

The potential of resource exploitation in the Antarctic has been a major *exogenous* driver for many years and was a key factor leading to the development of CAMLR Convention in the late 1970s. This convention was concluded in 1980 and entered into force in 1982 (Convention on the Conservation Antarctic Marine Living Resources (CCAMLR) 1980). CAMLR was sparked by concerns over unregulated exploitation of Southern Ocean krill, a keystone species in the marine food web of the region.

The mineral and hydrocarbon potential of Antarctica in the continental shelf areas surrounding the continent has been discussed for many decades. Despite robust assessments identifying constraints, challenges, the lack of proven prospects, as well as the current total prohibition on commercial mineral activity, the debate over mining persists. This reinforces the view of these economic driving forces as potential rather than actual, with regulatory arrangements acting as strong *constraints*.

3. **Social driving forces**

Antarctica has a special place in the human imagination. We have ongoing interest in the human struggles and sacrifice in the 'Heroic Era'—summed up by Apsley Cherry-Garrard (a member of Scott's fateful final expedition) as 'the worst journey in the world' (Cherry-Garrard 2010). An extreme and harsh environment—described by Australian expeditioner Douglas Mawson as 'the home of the blizzard' (Mawson 2015)—is a factor in the increasing attraction of the region as a tourist destination. Antarctic fauna, too, provides interest. Antarctica is a place apart (Walton 2016), epitomised by penguins as key images of the far south, whether appearing in nature documentaries (occasionally with their life cycle anthropomorphised) or being represented comically.

Another example of a social driver highlighted in the literature is collaboration between nations in common scientific endeavours. Assistance is willingly provided in cases of accidents or misfortune, for example. The Scientific Committee on Antarctic Research (SCAR)—itself a significant social driving force—has emphasised the worth of Antarctica to humankind: the time, effort, and money put into work on the continent, and the benefits that humankind receives from this investment (Scientific Committee on Antarctic Research 2017). Support for Antarctic scientific research is not simply an outcome of strategic or political drivers; it reflects commitments from ATCPs to maintain their obligations embedded in the Antarctic Treaty to ensure, as noted in theMadrid Protocol, that the region is maintained as ‘a natural reserve devoted to peace and science’ (Protocol on Environmental Protection to the Antarctic Treaty (Madrid Protocol) 1991, Art 2). This reinforces the importance of these principles as *endogenous* driving forces.

4. **Technological driving forces**
   Technology has long been an *impelling* driver of Antarctic endeavours. Advances in upper atmospheric physics following the Second World War were among the factors that encouraged the establishment of the International Geophysical Year (IGY) in 1957–58, and the specific attention on the Antarctic in this period. Since then, technology has revolutionised all aspects of Antarctic operations, from logistics and supply to science. Remote sensing and autonomous sensors provide data streams and access to locations unimagined in previous decades. Easier access to Antarctica, as well as enhanced communications and living conditions on the continent, have improved opportunities for high-quality science, a key for *cooperation* in the Antarctic. Revolutions in satellite-based navigation have increased access to high latitude areas, and remote sensing has provided tools for science to pursue big scientific questions.
   Technology may be ‘dual-use’, in the sense of being able to be seamlessly re-purposed between scientific and military capabilities (Brady 2017b; Flamm 2018) that may place pressure on the Antarctic Treaty’s prohibition of measures ‘of a military nature’ (Arts I and V) (Rothwell 1996). Such dual-use technology can include satellite receiving stations and infrared telescopes with capability to track satellites, missiles and drones (Brady 2017b). A recent paper by Bergin and Press for the Australian Strategic Policy Institute analyses this issue with respect to China, and recommends that increased inspections be carried out under the Antarctic Treaty to lessen the prospects of dual-use equipment being used for military purposes (Bergin and Press 2020). In 1990, Hemmings commented on the possibilities of ‘dual use’ of military equipment for both peaceful and military purposes in the ATA (Hemmings 1990). This theme has been explored most recently by Brady in her book, *China as a Great Polar Power* (2017) (Brady 2017b). Brady raises concern regarding Chinese Antarctic research bases hosting telecommunications equipment that might be used for military purposes, such as surveillance and missile targeting. This is examined further in Chap. 6.
5. **Environmental driving forces**
   Antarctica is known as the coldest, driest, highest and windiest continent—a place of environmental extremes. The Antarctic continent and the Southern Ocean is both an engine room for regional and global climate and a laboratory for research on climate and climate change (Haward 2019). A key challenge is human-induced climate change. This is affecting the Antarctic cryosphere and Southern Ocean marine ecosystems through increased warming and from the impacts of increased uptake of CO_2 in the ocean, leading to ocean acidification (Intergovernmental Panel on Climate Change (IPCC) 2019).

The impacts of climate change on Antarctica are contributing to ecological change in the region. Shifts in the range and distribution of species have been observed, along with changes in the seasonal extent and thickness of sea ice. Antarctica and the Southern Ocean are key components of the global climate system, and so these climatic alterations are likely to have broader impacts on ecosystems and primary production within the food web.

Climate change is also increasing access to high latitude areas (assisted by advances in technology) as sea ice extent diminishes. This opens up more areas for resource harvesting (for example, fisheries) and human access (Antarctic sea-borne tourism). As a result, climate change directly impacts environmental management in the Antarctic and Southern Ocean.

Notwithstanding the significance of Antarctica and the Southern Ocean in the global climate system, it was not until 2007, after 'extensive consideration', that the Antarctic Treaty Consultative Meeting (ATCM) XXX 'agreed to revise Agenda Item 13 to read Scientific Issues, including Climate Related Research, Scientific Cooperation and Facilitation' (Kelly and Child 1988). As a result, the ATCM has increased its attention to climate risk management (Comba 2013), recognising the salience of Antarctic science to address the critical *uncertainties* of climate change. These uncertainties include, for example, the extent to which changes in mass balances of Antarctic ice sheets have impacts on global mean sea levels.

6. **Legal driving forces**

We suggest that geopolitical scenarios for Antarctica need to scan for potential impacts from a broader set of legal fields. While the Antarctic Treaty and its linked instruments and institutions forming the ATS is a unique legal arrangement related to sovereignty, this does not mean that legal instruments do not shape and regulate behaviour. In addition to the Measure, Decisions and Resolutions passed at the ATCM or the Conservation Measures developed by CCAMLR, each ATCP has national laws related to Antarctica that create *certainty*, and, in turn, govern the behaviour of their citizens in Antarctica.

Outside the ATS, there are numerous other treaties, international instruments, bilateral and multilateral agreements, either with specific relevance to Antarctica, or with global application (Saul and Stephens 2015). These instruments include the 1946 International Convention for the Regulation of Whaling (ICRW); the 1972 World Heritage Convention; the 1974 Safety of Life at Sea Convention (SOLAS); the 1973/78 Convention for the Prevention of Pollution from Ships (MARPOL); the 1973 Convention on International Trade of Endangered Species Wild Flora and Flora (CITES); the 1972/1996 Convention on the Prevention of Marine Pollution by Dumping of Wastes and Other Matter (London Convention/Protocol): the 1979 Convention on Migratory Species of Wild Animals (CMS): the 1982 United Nations Convention on the Law of the Sea (UNCLOS); the 1992 Convention on Biological Diversity (CBD); and the 1992 United Nations Framework Convention on Climate Change (UNFCCC).

The ATS has closely guarded its claim to legal competence in the Antarctic and the Southern Ocean. Attempts by The Netherlands to encourage engagement with the Convention on Biological Diversity on biological prospecting were soundly rejected. Attempts by environmental non-governmental organisations to link Antarctica with the World Heritage Convention have been ignored. An attempt by Australia and New Zealand to link the Convention on International Trade of Endangered Species of Wild Flora and Flora (CITES) with CCAMLR measures on toothfish regulation was soundly defeated (Haward 2009). In the 2010s, however, the ATCM supported the adoption of the International Maritime Organization (IMO) standards for shipping under the International Code of Safety for Ships Operating in Polar Waters (Polar Code) (Jabour 2014).

7. **Geographic driving forces**

Antarctica's geography is a major influence on, and driver for, all activities in the region. As noted above, its extreme environment poses challenges for operations, including undertaking science—the *raison d'être* for Antarctic national programs. Antarctica's remote location drove 'Heroic Era' exploration and conquests, as it became the 'last continent' to fall to the great European imperial carve-up of the nineteenth century. Beck, among others, has noted that Antarctica's isolation also quarantined it from external tensions, and helped *constrain* destabilising rivalries that impacted other areas of the globe. As a result, the settlement provided by the Antarctic Treaty gave the region stability that, in general, proved to be of systemic benefit to the USA, Soviet Union and southern hemisphere countries (Beck 1990).

A consequence of the region's geographical isolation—except, of course, the relative proximity of the Antarctic Peninsula to Argentina and Chile—reinforces the classic geopolitical aphorism that 'geography is destiny' (Schwartz 2008). Antarctica's geography creates challenging logistics, high operating costs, and risk for countries operating in the region—and particularly for those working in East Antarctica. Antarctic activities are not cheap and need considerable advance planning—and can be upset by unanticipated changes in weather or incidents such as ships becoming trapped in sea ice.

Physically, the region covers approximately 20 per cent of the southern hemisphere (National Geographic Society 2021) and, as we have noted, it plays a major role in global climate. The Southern Ocean's circumpolar current is significant, as is the role of cold Antarctic bottom water that moves northwards, before rising as it warms. This contributes to the 'overturning circulation': the feature that moves ocean stores of heat from the tropics to the colder Southern Ocean. The dynamics of the physical geography of the region are thus a key driver or 'engine' of the global climate system (Intergovernmental Panel on Climate Change (IPCC) 2019).

8. **Strategic driving forces**

Strategic issues have long shaped interests in Antarctica. As the competition between the Soviet Union and the USA shaped the Cold War, their interests in Antarctica were hot issues for the future of the coldest continent. The Antarctic Conference, held in Washington, D.C. in 1958–59, led to the conclusion of the Antarctic Treaty, and as Hemmings, Dodds, and Roberts observe, involved only 12 states who designed 'the political framework for an entire continent'

(Hemmings et al. 2017, p. 2). The 12 states were those active in the IGY and included the USSR, despite internal debate within the US Government over whether to include the USSR.

The Antarctic Treaty has been very successful in ensuring that Antarctica and the Southern Ocean have remained free from conflict, retained its non-militarised status and avoided territorial disputes within the ATA. The Antarctic Treaty prohibits any measures of a military nature in Antarctica, such as 'the establishment of military bases and fortifications, the carrying out of military manoeuvres, as well as the testing of any type of weapons' (Art I)—see Chap. 6.

The ongoing success of the Antarctic Treaty's prohibition of nuclear weapons within the ATA is a notable achievement as the world's first nuclear weapons treaty (Nyman 2017; Abdenur and Marcondes 2014). The non-militarisation of Antarctica, including from nuclear weapons, has also performed a strategic role *outside* of Antarctica, in providing *certainty* and an initial southern example for other regional, nuclear weapon-free agreements (Herr et al. 1990).

As Young observes, the non-militarisation of Antarctica has had the effect of 'partially exempting it from the most vigorous and ongoing competition between the superpowers for power and influence' (Young 2017), p. xiv). He also notes, however, that the location and size of some research stations is driven by '[t]hinly disguised economic and political motivations' (Young 2017, p. xiv).

In 1990, just after the 30th anniversary of the signing of the Antarctic Treaty, the Australian Institute of International Affairs published an edited volume on the future of the ATS (Herr et al. 1990). This contained various contributions from experts on Antarctic affairs reviewing the history of Antarctic politics, including issues of sovereignty, conservation, environmental protection, mineral resource extraction and marine living resources. In a chapter titled 'Antarctica as a Continent for Science', Patrick Quilty, then Assistant Director, Science, at the Australian Antarctic Division, argued strongly that 'science is the currency of credibility' within the ATS (Quilty 1990). An interesting chapter by Peter Beck sits uneasily within the collection (Beck 1990). Titled 'Antarctica is a Zone of Peace: A Strategic Irrelevance?', Beck's chapter departs from the mainstream narrative regarding Antarctica is a zone for peace and science by raising issues relating to the contemporary strategic value of Antarctica from a traditional military point of view.

As we will discuss in Chap. 6, the Falklands/Malvinas war in 1982 brought military conflict between the United Kingdom and Argentina to the sub-Antarctic island areas of the South Atlantic. The sovereignty dispute between the UK and Argentina over the Falkland Islands/Islas Malvinas is, of course, outside the ATA (Benwell 2014, 2017). The overlapping sovereignty claims of Argentina, the UK and Chile within the ATA went to the heart of the 'Antarctic Problem' of the 1940s and 1950s—an issue largely resolved by the Antarctic Treaty.

In the context of increasing attention on the Indian Ocean, India includes Antarctica in 'expanding circles of engagement' over its maritime interests (Haward and Mason 2011, p. 62). This spatial framing (Powell and Dodds

2014) is not new. As Indian Prime Minister Indira Gandhi noted in 1983, 'The Indian Ocean links India to Antarctica. The entire area is of deep interest to us' (Chaturvedi 2013, p. 62). This reinforces Orakhelashvili's observation that geopolitics includes, among other things, 'explaining the need to access or control a remote territory, even if there is no immediate geographic link to it' (Orakhelashvili 2008, p. 161).

9. **Historical driving forces**
   In Chap. 2, we noted the significance of early Antarctic exploration on subsequent geopolitical developments. The 'Heroic Era' (1890s–1920s) provided the base for subsequent territorial claims and established the claimant states' interests in Antarctica (Haward 2017). This was reinforced by their role as key participants in the IGY of 1957–58 and the Antarctic Conference of 1958–59.

   We also noted that historical driving forces are not always positive. The critique of the 'closed shop' of Antarctic governance had its most significant support in the twenty years from the early 1980s, when Malaysia spearheaded a sustained critique of the ATS. This critique—couched as an item within the United Nations General Assembly—was primarily about equity of future access to Antarctic resources (Antarctica as the 'common heritage of mankind'). The 'Question of Antarctica' debates at the United Nations extended from 1982 to 2005 (Haward and Mason 2011) and, at the time, were interpreted as reflecting a 'crisis of legitimacy' for the ATS (Hemmings et al. 2017, p. 3). Malaysia's shift from critic to engaged participant in the ATS occurred in the early 2000s, culminating in its accession to the Antarctic Treaty in 2011. We should not underestimate the potential influence of these historical forces in contemporary and future events.

5.9 Summary

Our brief (and incomplete) classification of these driving forces highlights their many overlaps and interrelationships. We present them here as a snapshot of some of the key factors underpinning Antarctic geopolitics, which are likely to shape potential scenarios. It is clear that Antarctic geopolitics is marked by considerable complexity and uncertainty. We suggest these dynamics need to be understood and captured (at least to some degree). In particular, these driving forces tend to differ according to whether they are *predetermined elements* or *critical uncertainties*; whether they *impel* events or trends forward or *constrain* them; whether they instigate *rivalry* or promote *cooperation*; and whether they are *endogenous* or *exogenous* to the field in question. We also suggest that unpacking our driving forces in this way provides a more nuanced understanding of how they are likely to behave within a scenario.

In Chaps. 6, 7 and 8, we take the issue of militarisation as a case study to illustrate a technique for constructing scenarios of Antarctic geopolitics.

References

Abdel-Motaal, Doaa. 2016. *Antarctica: The Battle for the Seventh Continent*. Santa Barbara, CA: Praeger.

Abdenur, Adriane Erthal, and Neto D. Marcondes. 2014. Rising Powers and Antarctica: Brazil's Changing Interests. *The Polar Journal* 4 (1): 12–27.

Agustan, Djoko Karmiadji, Mega N. Putri, Oni B. Bintoro, Ira N. Djarot, Hasan Soleh, Aris Irawan, et al. 2019. Combining Delphi Study and Scenario Planning for Indonesia Research Priorities in Maritime Sector. In *IEEE International Conference on Aerospace Electronics and Remote Sensing Technology (ICARES)*, Yogyakarta, Indonesia, 17–18 October 2019, 1–7. New York: IEEE. https://ieeexplore.ieee.org/document/8914342. Accessed July 19, 2021.

Amer, Muhammed, Tugrul U. Daim, and Antonie Jetter. 2013. A Review of Scenario Planning. *Futures* 46: 23–40.

Jackson, Andrew, and Peter Boyce. 2011. Mining and 'Word Park Antarctica.' In *1982–1991 Australia and the Antarctic Treaty System: Fifty Years of Influence*, ed. Marcus Haward and Tom Griffiths, 243–273. Sydney: UNSW Press.

Augier, Mie, Nicholas Dew, Thorbjørn Knudsen, and Nils Stieglitz. 2018. Organizational Persistence in the Use of War Gaming and Scenario Planning. *Long Range Planning* 51 (4): 511–525.

Axelrod, Robert. M. 1997. *The Complexity of Cooperation: Agent-Based Models of Competition and Collaboration*. Princeton: Princeton University Press.

Beck, Peter J. 1990. Antarctica as a Zone of Peace: A Strategic Irrelevance? In *Antarctica's Future: Continuity or Change?* eds. Richard A. Herr, H. Robert Hall and Marcus G. Haward, 193–224. Hobart: Tasmanian Government Printer.

Benwell, Matthew C. 2014. Connecting Southern Frontiers: Argentina, the South West Atlantic and 'Argentine Antarctic Territory.' In *Polar Geopolitics: Knowledges, Resources and Legal Regimes*, ed. Richard C. Powell and Klaus Dodds, 201–218. Cheltenham, UK: Edward Elgar.

Benwell, Matthew C. 2017. Argentine Territorial Nationalism in the South Atlantic and Antarctica. In *Handbook on the Politics of Antarctica*, ed. Klaus Dodds, Alan D. Hemmings, and Peder Roberts, 540–551. Cheltenham, UK: Edward Elgar.

Bergin, Anthony, and Press, Tony. 2020. *Eyes Wide Open: Managing the Australia-China Antarctic Relationship*. ASPI Special Report 153, 27 April. Canberra: Australian Strategic Policy Institute.

Bradfield, Ronald, James Derbyshire, and George Wright. 2016. The Critical Role of History in Scenario Thinking: Augmenting Causal Analysis within the Intuitive Logics Scenario Development Methodology. *Futures* 77: 56–66.

Brady, Anne-Marie. 2013a. Introduction. Conflict or Cooperation? The Emerging Politics of Antarctica. In *The Emerging Politics of Antarctica*, ed. Anne-Marie Brady, 1–9. Abingdon, UK: Routledge.

Brady, Anne-Marie. 2013b. China's Antarctic Interests. In *The Emerging Politics of Antarctica*, ed. Anne-Marie. Brady, 31–49. Abingdon, UK: Routledge.

Brady, Anne-Marie. 2017a. The Past in the Present: Antarctica in China's National Narrative. In *Handbook on the Politics of Antarctica*, ed. Klaus Dodds, Alan D. Hemmings, and Peder Roberts, 284–300. Cheltenham, UK: Edward Elgar.

Brady, Anne-Marie. 2017b. *China as a Polar Great Power*. Cambridge: Cambridge University Press.

Cairns, George, Iftekhar Ahmed, Jane Mullett, and George Wright. 2013. Scenario Method and Stakeholder Engagement: Critical Reflections on a Climate Change Scenarios Case Study. *Technological Forecasting and Social Change* 80 (1): 1–10.

Chaturvedi, Sanjay. 2013. India and Antarctica: Towards Post-Colonial Engagement? In *The Emerging Politics of Antarctica*, ed. Anne-Marie. Brady, 50–74. Abingdon, UK: Routledge.

Cherry-Garrard, Apsley. 2010 [1922]. *The Worst Journey in the World*. London: Vintage Books.

Child, Jack. 1988. *Antarctica and South American Geopolitics: Frozen Lebensraum*. New York: Praeger.

Cohen, Saul B. 2014. *Geopolitics: The Geography of International Relations*. Lanham, Maryland: Rowman & Littlefield.

Collis, Christy. 2017. Territories Beyond Possession? Antarctica and Outer Space. *The Polar Journal* 7 (2): 287–302.

Comba, Danilo. 2013. An Analysis of the Impact of Climate Change for the Management of the Antarctic Region. In *Climate Change and Disaster Risk Management*, ed. Walter L. Filho, 311–319. Berlin: Springer.

Convention on the Conservation Antarctic Marine Living Resources (CCAMLR). Opened for signature 20 May 1980. 1329 UNTS 47 (entered into force 7 April 1982).

Derbyshire, James, and George Wright. 2017. Augmenting the Intuitive Logics Scenario Planning Method for a More Comprehensive Analysis of Causation. *International Journal of Forecasting* 33 (1): 254–266.

Dodds, Klaus, and Mark Nutall. 2016. *The Scramble for the Poles: The Geopolitics of the Arctic and Antarctic*. Cambridge: Polity Press.

Flamm, Patrick. 2018. Geopolitics and New Zealand's Antarctic Presence. *The Diplomat*, 25 July. Washington, DC: Diplomat Media Inc. https://thediplomat.com/2018/07/geopolitics-and-new-zealands-antarctic-presence/. Accessed 10 October 2019.

Flint, Colin, and Virginie Mamadouh. 2015. The Multi-Disciplinary Reclamation of Geopolitics: New Opportunities and Challenges. *Geopolitics* 20 (1): 1–3.

Haward, Marcus. 2009. The Law of the Sea Convention and the Antarctic Treaty System: Constraints or Complementarity. In *Maritime Boundary Disputes, Settlement Processes, and the Law of the Sea*, eds. Seoung-Yong Hong and Jon. M. van Dyke, 231–251. Leiden: Martinus Nijhoff Publishers.

Haward, Marcus. 2017. The Originals: The Role and Influence of the Original Signatories to the Antarctic Treaty. In *Handbook on the Politics of Antarctica*, ed. Klaus Dodds, Alan D. Hemmings, and Peder Roberts, 231–240. Cheltenham, UK: Edward Elgar.

Haward, Marcus. 2019. Climate Change and the Southern Ocean: The Regime Complex for Regional Governance. In *Climate Change and Ocean Governance: Politics and Policy for Threatened Seas*, ed. Paul G. Harris, 201–214. Cambridge: Cambridge University Press.

Haward, Marcus, and David Mason. 2011. Australia, the United Nations and the Question of Antarctica. In *Australia and the Antarctic Treaty System: Fifty Years of Influence*, ed. Marcus Haward and Tom Griffiths, 202–221. Sydney: UNSW Press.

Hemmings, Alan. D. 2012. Considerable values in Antarctica. *The Polar Journal* 2(1): 139–156.

Hemmings, Alan D. 1990. Is Antarctica Demilitarised? In *Antarctica's Future: Continuity or Change?* eds. Richard A. Herr, H. Robert Hall and Marcus G. Haward, 225–241. Hobart: Tasmanian Government Printer.

Hemmings, Alan D. 2017. Antarctic politics in a transforming global geopolitics. In *Handbook on the Politics of Antarctica*, ed. Klaus Dodds, Alan D. Hemmings, and Peder Roberts, 507–522. Cheltenham, UK: Edward Elgar.

Hemmings, Alan D., Klaus Dodds, and Peder Roberts. 2017. Introduction: The Politics of Antarctica. In *Handbook on the Politics of Antarctica*, ed. Klaus Dodds, Alan D. Hemmings, and Peder Roberts, 1–17. Cheltenham, UK: Edward Elgar.

Herr, Richard A., H. Robert Hall, and Marcus G. Haward. 1990. *Antarctica's Future: Continuity or Change?* Hobart: Tasmanian Government Printer.

Huber, Jan. 2011. The Antarctic Treaty: Toward a New Partnership. In *Science Diplomacy: Antarctica, Science, and the Governance of International Spaces,* ed. Paul A. Berkman, Michael A. Lang, David W.H. Walton and Oran R. Young, 89-95. Washington, D.C.: Smithsonian Institution Scholarly Press.

Intergovernmental Panel on Climate Change (IPCC) 2019. *IPCC Special Report on the Ocean and Cryosphere in a Changing Climate.* Hans-Otto Pörtner, Debra C. Roberts, Valérie Masson-Delmotte, Panmao Zhai, Melinda Tignor, Elvira Poloczanska, Katja Mintenbeck, et al. eds. https://www.ipcc.ch/ srocc. Accessed 10 January 2021.

Jabour, Julia. 2014. Progress Towards the Mandatory Code for Polar Shipping. *Australian Journal of Maritime and Ocean Affairs* 6 (1): 64–67.

Jones, Alun, and Julian Clark. 2019. Performance, Emotions, and Diplomacy in the United Nations Assemblage in New York. *Annals of the American Association of Geographers* 109 (4): 1262–1278.

Kaye, Stuart, Michael Johnson, and Rachel Baird. 2011. Law. In *Australia and the Antarctic Treaty System: Fifty Years of Influence*, ed. Marcus Haward and Tom Griffiths, 97–117. Sydney: UNSW Press.

Kelly, Philip, and Jack Child, eds. 1988. *Geopolitics of the Southern Cone and Antarctica.* Boulder, Colorado: Lynne Rienner Publishers.

Kelly, Phil. 2016. *Classical Geopolitics: A New Analytical Model.* Stanford: Stanford University Press.

Liggett, Daniela. 2015. An Erosion of Confidence. In *Diplomacy on Ice: Energy and the Environment in the Arctic and Antarctic*, ed. Rebecca Pincus and Saleem H. Ali, 61–71. New Haven: Yale University Press.

Liggett, Daniela, Bob Frame, Neil Gilbert, and Fraser Morgan. 2017. Is it All Going South? Four Future Scenarios for Antarctica. *Polar Record* 53 (5): 459–478.

Mackinder, H.J. 1904. The Geographical Pivot of History. *The Geographical Journal* 23 (4): 421–437.

Mahan, Alfred T. 2010a [1890]. *The Influence of Sea Power Upon History, 1660–1783.* Cambridge Library Collection—Naval and Military History. Cambridge: Cambridge University Press.

Mahan, Alfred T. 2010b [1892]. *The Influence of Sea Power upon the French Revolution and Empire, 1793–1812*, Vol. 1. Cambridge Library Collection - Naval and Military History. Cambridge: Cambridge University Press.

Martin, Lawrence. 1949. The Antarctic Sphere of Interest. In *New Compass of the World: A Symposium on Political Geography*, eds. Hans W. Weigert, Vilhjalmur Stefansson and Richard Edes Harrison, 61–79. New York: The Macmillan Company.

Mawson, Douglas. 2015 [1915]. *The Home of the Blizzard: Being the Story of the Australasian Antarctic Expedition 1911–1914.* Hobart: Australian Antarctic Division. https://www.antarctica.gov.au/site/assets/files/47954/home-of-the-blizzard.pdf. Accessed July 26, 2021.

Meissner, Philip, Christian Brands, and Torsten Wulf. 2017. Quantifying Blind Spots and Weak Signals in Executive Judgment: A Structured Integration of Expert Judgment into the Scenario Development Process. *International Journal of Forecasting* 33 (1): 244–253.

National Geographic Society. 2021. *Antarctica.* Resource Library, Encyclopaedic Entry. Washington, DC: National Geographic Society. https://www.nationalgeographic.org/encyclopedia/antarctica/. Accessed 26 July 2021.

Nyman, Elizabeth. 2017. Contemporary Security Concerns. In *Handbook on the Politics of Antarctica*, ed. Klaus Dodds, Alan D. Hemmings, and Peder Roberts, 571–583. Cheltenham, UK: Edward Elgar.

Orakhelashvili, Alexander. 2008. International Law and Geopolitics: One Object, Conflicting Legitimacies? *Netherlands Yearbook of International Law* 39: 155–204.

Oxford English Dictionary, [Online], 2021. Oxford: Oxford University Press. https://www.oed.com. Accessed July 25, 2021.

Powell, Richard C., and Klaus Dodds. 2014. Polar Geopolitics. In *Polar Geopolitics: Knowledges, Resources and Legal Regimes*, ed. Richard C. Powell and Klaus Dodds, 3–18. Cheltenham, UK: Edward Elgar.

Proctor, Robert W., and Kim-Phuong L. Vu. 2019. How Psychologists Help Solve Real-World Problems in Multidisciplinary Research Teams. *American Psychologist* 74(3): 271–277.

Protocol on Environmental Protection to the Antarctic Treaty (Madrid Protocol). Opened for signature 4 October 1991. 30 ILM 1455 (entered into force 14 January 1998).

Quilty, Patrick G. 1990. Antarctica as a Continent for Science. In *Antarctica's Future: Continuity or Change?* ed. Richard A Herr, H. Robert Hall and Marcus G. Haward, 29–37. Hobart: Tasmanian Government Printer.

Ramírez, Rafael, and Angela Wilkinson. 2016. *Strategic Reframing: The Oxford Scenario Planning Approach*. Oxford: Oxford University Press.
Rothwell, Donald. 1996. *The Polar Regions and the Development of International Law*. Cambridge: Cambridge University Press.
Saul, Ben, and Tim Stephens. 2015. *Antarctica in International Law*. Oxford: Hart Publishing.
Schwartz, Peter. 1996. *The Art of the Long View: Planning for the Future in an Uncertain World*. Chichester, UK: Wiley & Sons.
Schwartz, Benjamin. 2008. Geography is Destiny. *The Atlantic*, December 2008. https://www.theatlantic.com/magazine/archive/2008/12/geography-is-destiny/307163/. Accessed 26 July 2021.
Scientific Committee on Antarctic Research. 2017. *Strategic Plan 2017–2022: Connecting and Building Antarctic Research*. Cambridge: SCAR. https://www.scar.org/strategic-plans/774-2017-strategic-plan/. Accessed 26 July 2021.
Sloan, Geoffrey. 2017. *Geopolitics, Geography and Strategic History*. Abingdon, UK: Routledge.
Spykman, Nicholas J., and Helen R. Nicholl, eds. 1944. *The Geography of the Peace*. New York: Harcourt, Brace and Company.
Steig, Eric. J. 2019. How Fast Will the Antarctic Ice Sheet Retreat? *Science* 364(6444): 936–937.
Stokke, Olaf, and Davor Vidas, eds. 1996. *Governing the Antarctic: The Effectiveness and Legitimacy of the Antarctic Treaty System*. Cambridge: Cambridge University Press.
United States Congress, Office of Technology Assessment. 1989. *Polar Prospects: A Minerals Treaty for Antarctica OTA-O-428*. Washington, DC: US Government Printing Office.
US Geological Survey, National Minerals Information Center. 2021. *Rare Earths Statistics and Information*. Reston, Virginia: USGS. https://www.usgs.gov/centers/nmic/rare-earths-statistics-and-information. Accessed July 26, 2021.
Van der Heijden, Kees. 1996. *Scenarios: The Art of Strategic Conversation*. Chichester, UK: Wiley & Sons.
Van Efferink, Leonhardt. 2017. Geopolitics. In *Oxford Bibliographies*. New York: Oxford University Press.
Walton, Rowana. 2016. The Frozen White Continent—A Place Apart. *WiseOceans Blogs*. Perth and Kinross, UK: WiseOceans. https://www.wiseoceans.com/the-frozen-white-continent-a-place-apart/. Accessed July 26, 2021.
Wehrmann, Dorothea. 2019. *Critical Geopolitics of the Polar Regions: An Inter-American Perspective*. Abingdon, UK: Routledge.
Wright, George, Ron Bradfield, and George Cairns. 2013. Does the Intuitive Logics Method—and Its Recent Enhancements—Produce 'Effective' Scenarios? *Technological Forecasting and Social Change* 80 (4): 631–642.
Young, Oran R. 2010. *Institutional Dynamics: Emergent Patterns in International Environmental Governance*. Cambridge, MA: The MIT Press.
Young, Oran R. 2017. Foreword: Why Should We Take an Interest in What Happens in Antarctica? In *Handbook on the Politics of Antarctica*, ed. Klaus Dodds, Alan D. Hemmings, and Peder Roberts, xiv–xvi. Cheltenham, UK: Edward Elgar.

Chapter 6
Militarisation of Antarctica

Abstract In this chapter, we settle on the topical issue of militarisation for our scenarios. Antarctica is commonly perceived as immune from military competition. However, prior to the Antarctic Treaty, there was a small but significant history of military activity and skirmishes. The Antarctic Treaty has ensured that the ATA remains free from conflict, retains its non-militarised status, and has avoided the escalation of territorial disputes. The non-militarisation of Antarctica also performs a strategic role outside Antarctica, in providing a stable, southern buttress for other regional agreements that ban nuclear weapons. Antarctica's geographic isolation and severe climate present huge costs, logistical barriers, and risks for military activity on the continent—particularly in East Antarctica. However, some technologies being deployed within the ATA may be 'dual use', in the sense of being able to be seamlessly re-purposed between scientific and military capabilities. The greatest risk here is the potential for dual-use equipment and personnel in the ATA to be re-purposed to support non-peaceful activities *outside* the ATA—on land, in the oceans, or in space. The modest frequency of compliance inspections and the need for highly technical expertise to detect breaches mean that prospects of policing this activity effectively are slight.

Keywords Militarisation of Antarctica and the Southern Ocean · Antarctic militarisation history · Antarctic Treaty as a peace treaty · Non-militarisation of Antarctica · Strategic interests in Antarctica · Dual-use equipment and personnel in Antarctica · Antarctic inspection regime

6.1 Introduction

Antarctica is commonly perceived (at least in the public imagination) as an area of the planet which has been largely immune from the pressures of military competition. This perception stems partly from the remote and inaccessible geography of the Antarctic region. It is also a testament to the remarkable success of the 1959 Antarctic Treaty. The Treaty has managed international discord over territorial claims and avoided militarisation of the area below 60° South—which amounts to approximately 6.75 % of the surface area of the planet. The 1991 Protocol on Environment Protection

J. McGee et al., *The Future of Antarctica*, Springer Polar Sciences,
https://doi.org/10.1007/978-981-16-7095-4_6

to the Antarctic Treaty (Madrid Protocol) reinforced these principles of peace, science and absence of military competition by designating Antarctica as a 'natural reserve devoted to peace and science' (Art 2).

If we consider only the last 60 years of Antarctic history, we could be excused for concluding that the militarisation of Antarctica is an extremely low probability. However, the earlier history of the Antarctic continent and surrounding Southern Ocean was not free of military competition. Indeed, formation of the Antarctic Treaty was driven significantly by concerns over potential militarisation of the continent. A treaty was necessary if Antarctica were to be a zone free from state conflict, and a place where international scientific cooperation could proceed.

This chapter provides background to the issue of militarisation (and non-militarisation) in Antarctica and the Southern Ocean and, in particular, the question of 'dual-use' equipment and personnel. It provides the foundation for our case study in Chaps. 7 and 8, where we use the scenario planning approach discussed in Chaps. 3–5 to develop five scenarios relating to militarisation. The chapter proceeds as follows. First, we provide an overview of the history of military and strategic issues in the area below 60° South—the Antarctic Treaty Area (ATA), including important historical background from the Second World War and early Cold War years. Next, we describe how Article I of the Antarctic Treaty resolved military tensions by prohibiting state measures of a military nature being carried out within the ATA. We also explain the permitted roles of military personnel and equipment in carrying out activities in Antarctica through logistical support for scientific research, and for other peaceful purposes. Then, looking to the present day and to the immediate future, we analyse current military tensions that might impact upon the ATA. We expand this analysis to examine the issue of 'dual-use' equipment and personnel, and the inspection processes provided by the Antarctic Treaty and Madrid Protocol. We conclude by anticipating how this chapter will inform the scenario analysis that follows in Chaps. 7 and 8.

6.2 History of Militarisation in Antarctica

The largely serene history of Antarctica in the 30 years since the adoption of the Madrid Protocol tends to mask the mood of earlier periods, in which Antarctica and the Southern Ocean were sites of significant international discord. A key theatre of this earlier tension was the Antarctic Peninsula, with its three overlapping territorial claims of the United Kingdom (1908), Chile (1940) and Argentina (1942). These claims have been the source of enduring friction over the Antarctic Peninsula (see Chap. 2). Argentina and Chile contend that their respective territorial claims encompass areas of Antarctica that were originally dedicated to Spain by Papal Bulls (declarations) stretching back to 1493. Argentina and Chile also assert that these areas were inherited by them from Spain upon decolonisation, and that subsequently their legal titles have been bolstered through effective national occupation. Argentina therefore considers the area of its claim as a long-held Argentinian territory, the legal title to which predates the British claims of 1908. In the view of some, Argentina was late to the party in formalising its claim to this area in the mid-1940s; however, Argentina believed there was no necessity to do this for an area it already considered to be its territory. Adding further complexity, Chile also asserts a claim in this region. The United States and Soviet Union also individually asserted that they each had 'basis of claim' (that is, legal right to make a territorial claim) which had not yet been exercised. Although the issue of territorial claims on the Antarctic Peninsula has not been settled permanently, the Antarctic Treaty protected the status quo as at its entry into force in 1961. To that extent, the Treaty practically 'resolves' the issue, at least amongst its parties, by protecting different national positions and by setting aside disruptive arguments. Nonetheless, the question of claims in the Antarctic Peninsular remains an ongoing point of diplomatic tension between Argentina, Chile and the UK (Dodds 2002).

Germany has a significant historic connection with the Antarctic continent and Southern Ocean. There were three twentieth-century German Antarctic expeditions prior to the Second World War. The first (1901–03), led by geologist Erich von Drygalski in the *Gauss*, visited the Kerguelen Islands and Davis Sea area of East Antarctica (Rack et al. 2015). The second (1911–13), led by Wilhelm Filchner in the *Deutschland*, sailed to South Georgia and the Weddell Sea. The third (1937–38), led by Naval Captain Alfred Ritscher in the *Schwabenland*, explored areas of Queen Maud Land, both overland and via aerial photographic surveys. This Nazi-era expedition was tasked with finding a suitable area for a whaling station to increase Germany's supply of fats and oils (Lüdecke 2012).

During the Second World War (1939–45), Germany operated several 'surface raider' ships to target Allied supply lines—including in the Southern Ocean. In January 1941, the German surface raider *Pinguin* intercepted a Norwegian whaling fleet operating at 59° South in the Atlantic sector (Blyth 1952). In this incident, the *Pinguin* captured three Norwegian whaling ships and over 23,000 tonnes of whale oil. During the War, the Kerguelen Islands were used as a rendezvous and resting point for German raiders operating in the Atlantic and Indian areas of the Southern Ocean (Blyth 1952). The Kerguelen Islands were also the target of unrealised German plans to establish a meteorological and radio station (Blyth 1952). In 1941, this prompted the Royal Australian Navy to lay mines through the islands' strategic waters (Blyth 1952).

German wartime activity in the Southern Ocean (including the action by the *Pinguin*) spurred important developments in Antarctic history. First, in early 1941, Britain responded to the *Pinguin* incident by sending the armed merchant cruiser *Queen of Bermuda* to patrol the area between South Georgia, the South Shetland Islands and Weddell Sea in search of German raiders (Haddelsey and Carrol 2014, pp. 15–19). On 5 March 1941, the *Queen of Bermuda* visited an abandoned Norwegian whaling station on Deception Island and sabotaged the facility, so that it could not be used by Germany. In 1942, Argentina (then neutral) sent the transport vessel ARA *Primero de Mayo* to Deception Island, Melchior Islands, Palmer Archipelago and Winter Island to raise Argentinian flags and claim annexation of these areas below 60° South. Then, in early 1943, the British armed merchant cruiser *Carnarvon Castle* visited Deception Island and removed the Argentinian flags, replacing them with the British flag. A month later, the *Primero de Mayo* returned and reinstated the Argentinian flags. In 1944, Britain carried out a secret Antarctic expedition, 'Operation Tabarin', which involved fourteen men landing to establish a small civilian scientific presence on Deception Island, with permanent research bases (Haddelsey and Carrol 2014; Dudeney and Walton 2012). Thus, Germany's wartime activities near the Antarctic Peninsula brought to a head the longstanding territorial dispute between the UK and Argentina.

In the immediate post-war period, the first significant use of military capabilities in Antarctica was by the United States. By the mid-1940s, the US had not made a claim to Antarctic territory—despite three earlier Antarctic expeditions involving Admiral Richard E. Byrd (1928–30, 1934, 1938–40), as well as the aviation exploration activities of civilian Lincoln Ellsworth (including the first trans-Antarctic flight from Dundee Island to the Ross Ice Shelf in 1935). In 1946–47, however, the US used Antarctica for a major military training exercise when it embarked on 'Operation Highjump'—again led by Admiral Byrd—involving 15 US Navy ships, 29 aircraft and 4700 servicemen deployed to Marie Byrd Land and the Ross Sea. The objective of 'Operation Highjump' was to train US military forces in polar combat (primarily to thwart any Soviet invasion of the Arctic) and to undertake scientific investigation. 'Operation Highjump' reminds us that large-scale military activity has occurred in Antarctica in the relatively recent past. The US Navy also carried out further smaller expeditions to Antarctica: 'Operation Windmill' (1947–48) which involved two icebreakers; and 'Operation Deepfreeze I' (1956–57), in which seven freighters and tankers were deployed to support US contributions to the International Geophysical Year (IGY) of 1957–58.

The availability of post-war military equipment and the unique ability of military forces to operate in difficult conditions supported the expansion of Antarctic research, including during the Antarctic component of the IGY. Indeed, without this logistical support through use of the military, the IGY may not have been possible. For this reason, the Antarctic Treaty negotiators were keen to allow the continuing use of military facilities to support science. At least in the early years of the Antarctic Treaty, scientific research depended significantly on access to military resources. To a large degree, it still does. Had access to military facilities not been available, it is quite likely that the rapid expansion of Antarctic science would not have been possible.

Tensions between the UK and Argentina over their overlapping claims on the Antarctic Peninsula led to the only known military action on the Antarctic continent itself. In February 1952, at Hope Bay on the Antarctic Peninsula, the British research vessel *John Biscoe* attempted to unload equipment and stores to rebuild a fire-damaged British base. However, it was warned off by a landing party of Argentinian military, who fired machine guns above the heads of the British crew. Without Foreign Office approval, the Falkland Islands Governor, Sir Miles Clifford, returned with a British frigate and an accompanying detachment of Royal Marines, to escort the *John Biscoe* back to Hope Bay and provide protection whilst the British base was rebuilt. Peaceful relations between the UK and Argentina were soon restored.

While the Hope Bay incident was an inconsequential fracas, it shows that Antarctica has not been completely free from the use of military force to support territorial claims. If there was an enduring consequence, it was to emphasise the benefits of the non-militarisation pact in the Antarctic Treaty, which was to follow a few years later.

The following year, 1953, saw the UK exercise minor military force on Deception Island the South Shetland Islands—although this time no shots were fired. On 15 February, 35 Royal Marines arrested and deported two sailors from the Argentinian naval tug *Chiriguano*, which was positioned near the UK operated airstrip on Deception Island (Royal Institute of International Affairs 1953). A hut constructed in the area by these Argentinian sailors was dismantled. A second unoccupied hut in the area constructed by Chilean naval personnel was similarly dismantled by the Royal Marines (Royal Institute of International Affairs 1953). Argentina and Chile were informed of these actions by UK diplomatic correspondence; they replied by protesting and asserting sovereignty to the relevant area (Royal Institute of International Affairs 1953).

Figure 6.1, below, provides a map of known sites on military activities in the Antarctic region in the period leading up to formation of the Antarctic Treaty 1959.

During the post-war period of the late 1940s and 1950s, the potential for unresolved territorial claims in Antarctica to cause international conflicts (either directly, or as proxy disputes between the Cold War powers) was well recognised by the US and other key actors in the international system. The Korean Peninsula was a stark example of how Cold War geopolitical tensions could slide into military conflict. Tensions over territorial claims on the Antarctic Peninsula, and the potential for geopolitical rivalry between the US and the Soviet Union to spill into this theatre, led to the phrase the 'Antarctic Problem' being used frequently in diplomatic discourse. At the time, 'problem' was a common descriptor for tricky issues requiring diplomatic attention (Christie and Hunter 1951; Hall 1994). The US was aware of the conflict potential of the Antarctic Peninsula and was keen was keen to see an international solution (Hall 1994).

In the 1950s, heightened Cold War tensions were an important political backdrop for potential military use of Antarctica. During this period, the Soviet Union constructed year-round scientific research bases in East Antarctica. In preparation for research activities as part of the IGY, the Soviets established Mirny base in 1956 and Vostok base in 1957. A permanent Soviet presence in East Antarctica was of concern to the Australian Government, as these bases were being constructed within territory already claimed by Australia. Australia also held concerns that the Soviet IGY research bases may be the start of an ongoing Soviet presence in East Antarctica. This could lead to the basing of nuclear or conventional missiles, or other long-range weapons which might threaten southern areas of the Australian continent (i.e.,Tasmania, Victoria, South Australia, and Western Australia). Thus, the Australian Government had strong reasons to support the proposal that Antarctica and surrounding oceans be used for peaceful purposes only.

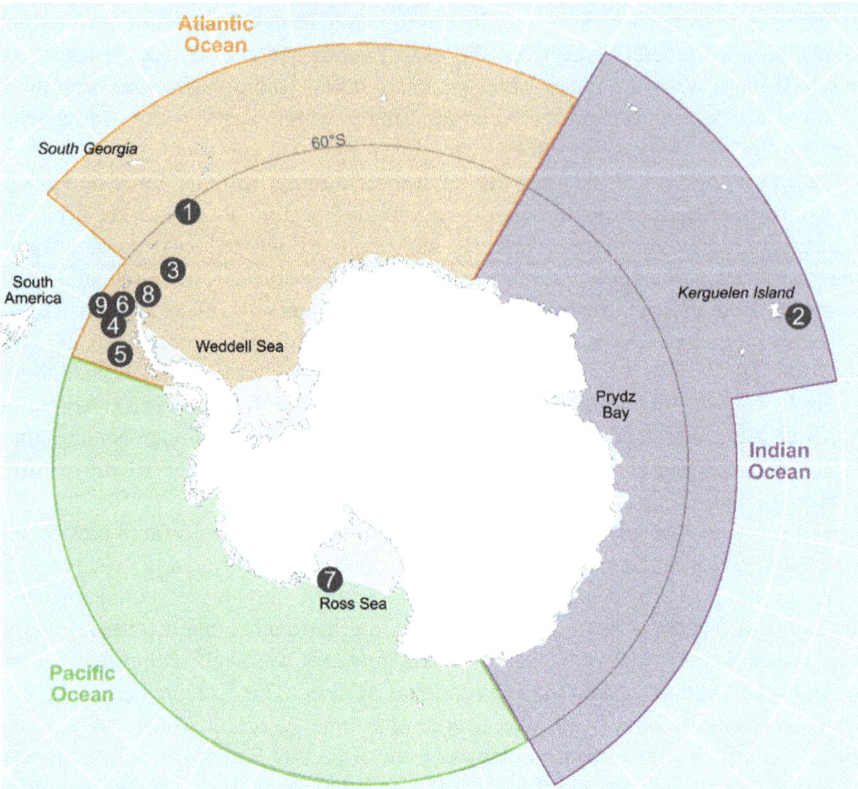

KEY

1. In 1941, German surface raider *Pinguin* captured 3 Norwegian whaling ships and over 23,000 tonnes of whale oil.
2. Rendezvous and resting point for German raider ships operating in the Atlantic and Indian areas.
3. In early 1941, the British send the armed merchant cruiser *Queen of Bermuda* to patrol the area between South Georgia, the South Shetland Islands and Weddell Sea in search of German raiders.
4. In March 1941, the *Queen of Bermuda* visited an abandoned Norwegian whaling station on Deception Island and sabotaged the facility, so that it could not be used by Germany.
5. In 1942, an Argentine Antarctic expedition took formal possession on behalf of the Argentine Republic of a sector of the Antarctic continent by raising Argentine flags and leaving certificates in the Deception Island, Melchior Island, and Argentine Islands.
6. In 1943, the British armed merchant cruiser *Carnarvon Castle* visited Deception Island and removed the Argentina flags, replacing them with the British flag. A month later the ARA *Primero de Mayo* returned and reinstated the Argentinian flags.
7. 1946-47, United States naval expedition 'Operation Highjump' in the Ross Sea.
8. In 1952 at Hope Bay, the British research vessel *John Biscoe* attempted to unload equipment and stores to rebuild a fire-damaged British base, but was warned off by a landing party of Argentinian military, who fired machine guns above the heads of the British crew.
9. In 1953, 35 armed British Royal Marines landed on Deception Island, capturing two sailors from Argentina, and destroying two small huts built by Argentina and Chile.

Fig. 6.1 Sites of Military Activity in the Antarctic Region prior to 1959

6.3 Managing Militarisation Under the Antarctic Treaty 1959

In 1959, after the IGY was completed, the United States took the lead in organising an international conference, held in Washington DC, to resolve the 'Antarctic Problem' and to reduce the potential for international conflict. The international treaty that was agreed at the Washington Conference—the Antarctic Treaty 1959—should therefore be viewed primarily as a peace treaty.

In the first substantive paragraph of the preamble, the 12 signatory countries recognise that:

> it is in the interest of all mankind that Antarctica shall continue forever to be used exclusively for peaceful purposes and shall not become the scene or object of international discord.

Similarly, Article I(1) states:

> Antarctica shall be used for peaceful purposes only. There shall be prohibited, inter alia, any measures of a military nature, such as the establishment of military bases and fortifications, the carrying out of military manoeuvres, as well as the testing of any type of weapons.

Article I(1) introduces a blanket obligation that Antarctica shall be used only for peaceful purposes and it prohibits any conduct that departs from this standard. Military assets (such as ships and aircraft) and personnel remain significant components of national Antarctic programs, and these may operate in Antarctica as declared support for a party's research and logistic activities. It also provides examples of such departures in a non-exhaustive list of 'measures of a military nature' (see above). These examples are consistent with military technologies that might have been anticipated in late 1950s. Later, we discuss the modern context of 'measures of a military nature' and argue that the concept has broadened from what was conceived when the Treaty was signed.

Three other articles of the Antarctic Treaty are also relevant to militarisation. First, the presence of military personnel on the Antarctic continent is expressly provided for. Article I(2) indicates the Treaty 'shall not prevent the use of military personnel or equipment for scientific research or for other peaceful purpose'. The Treaty clearly provides for military personnel being in Antarctica, for purposes such as providing logistical support for scientific activities. This was intended to continue the measures in place during the IGY, which had proved so effective. The ability to use military resources to support civilian research allowed states to pursue Antarctic programs without having to invest in additional logistic capabilities.

Second, Article V provides that 'any nuclear explosions in Antarctica and the disposal there of radioactive waste material shall be prohibited'. Importantly, the Antarctic Treaty is the first example of international law being used to prohibit the use of nuclear explosions in any area of the planet. The Treaty should be viewed as a seminal arms control treaty of the Cold War era. The prohibition on nuclear explosions in Article V extends to any civilian use of nuclear explosions, as there was some concern during negotiation of the Treaty that 'peaceful' nuclear explosions might be used for large scale engineering, such as removing ice sheets to enable mining.

Third, another key provision is the Treaty's pioneering inspection regime, provided in Article VII and reinforced later by Article 14 of the Madrid Protocol. This gives parties complete freedom of access at any time to all research stations, ships, and aircraft in Antarctica. Article VII(5)(c) of the Antarctic Treaty requires parties to give advanced notice of 'any military personnel or equipment intended to be introduced' into Antarctica for scientific research or other peaceful purposes. Such notice was intended to provide transparency and to allow for confirmation by inspection teams from other parties. Under this provision, Antarctic states are required to routinely disclose their planned uses of military personnel and equipment, in support of peaceful activities, for the upcoming research season.

Due to its geographical location, Antarctica has another important role in international arms control activity. The 1996 Comprehensive Nuclear Test Ban Treaty (CTBT) has been ratified by 168 states and prohibits states from detonating nuclear explosions, whether on the Earth's surface, in the atmosphere, in the oceans or underground. The CTBT makes it very difficult for states that are party to it to develop nuclear weapons for the first time, or to enhance their weapons stockpile with versions that are more powerful or effective. However, only 36 of the 44 nuclear technology states have signed and ratified the CTBT, and the Treaty requires signature and ratification by all 44 of these nuclear-capable states before it comes into effect. Despite this, the CTBT has developed an extensive international monitoring system (IMS) to encourage states to adhere to the prohibition on nuclear explosions. The IMS consists of 337 monitoring facilities across the planet that rely on seismic, hydroacoustic, infra-sound and radionuclide detection technologies. The Antarctic continent hosts 12 of these monitoring stations for the IMS (four seismic, four infra-sound and four radionuclide) located among US, German, French and Australian Antarctic research stations. The Antarctic continent's role in monitoring for nuclear explosions (whether inside or outside the ATA) is clearly a peaceful use of the continent and is consistent with the Antarctic Treaty's prohibition of nuclear explosions.

6.4 Traditional Military/Strategic Threats in Antarctica

Since the Antarctic Treaty came into force, the greatest threat of international conflict spilling into the ATA arose during the Falkland Islands / Islas Malvinas armed conflict between the UK and Argentina. This 10-week period of undeclared war in the South Atlantic commenced in early 1982. The conflict centred on three island groups that are subject to an ongoing territorial dispute between the UK and Argentina: the Falkland Islands (known in Argentina as 'Islas Malvinas'); South Georgia Island; and the South Sandwich Islands. South Georgia Island and the South Sandwich Islands are located close to the boundary of the ATA. The conflict began on 2 April 1982, when the Argentine military junta sent military forces to occupy the Falklands/Malvinas, which were under British administration. The longstanding territorial dispute over sovereignty of these islands also has a historical connection with the overlapping territorial claims of Argentina and the UK in the Antarctic Peninsula.

On 2 April 1982, Argentinian forces occupied the Falkland Islands/Islas Malvinas and South Georgia. The UK responded by dispatching a Royal Navy task force, including aircraft carriers, which engaged in a 74-day conflict with Argentinian military forces. On 2 May, the Argentinian battleship ARA *General Belgrano* was sunk by Royal Navy submarine HMS *Conqueror* at 55° South, with significant loss of life. Despite these hostile military operations relatively close to the ATA, there was no evidence of Argentine or UK forces breaching 60° South during these hostilities. The UK campaign was ultimately successful, with Argentine forces surrendering on 14 June and returning control of the islands to the UK. While this conflict occurred in the South Atlantic, the Falklands / Malvinas, South Georgia and South Sandwich Islands are all located north of the ATA. However, it is known that British Antarctic stations in the region (and Argentine stations) were used to transmit military communications during the conflict.

It is noteworthy that, during the periods before and after this military engagement, Argentina and the UK continued to participate in the Antarctic Treaty Consultative Meetings (ATCMs). Even at the height of the conflict in 1982, Argentine and British diplomats continued negotiating (along with the rest of the Antarctic Treaty parties) the problematic minerals convention, consciously quarantining external conflicts from Antarctic Treaty forums.

Aside from the Falklands / Malvinas war, there has been no serious threat of military hostilities between states spilling into in the ATA during the 60 years since the Antarctic Treaty came into force in 1961. Overall, the Treaty has been remarkably successful in maintaining non-militarisation of the area below 60° South. The territorial claims of the seven claimant states were not abandoned, but instead protected by the operation of Article IV(1). The position of countries asserting a basis of claim was similarly protected. This has proved quite successful in managing international discord over the territorial claims. Ongoing international tensions remain between Argentina, Chile, and the UK over their overlapping claims on the Antarctic Peninsula. However, this tension has been managed effectively through Article IV, which allows each state—regardless of its attitude to the territorial claims—to have its position protected, so that priority is given to the peaceful use and scientific investigation of Antarctica. The main reminder of the differences between Argentina and the UK at Antarctic Treaty meetings is Argentina's routine objection to the use of 'Falkland Islands' to refer to 'Islas Malvinas', as this might imply endorsement of UK sovereignty over the islands. The standard UK response is to assert confidence in its position, after which discussion resumes with the status quo upheld.

Importantly, the most significant challenge to the Antarctic Treaty System (ATS) in the 1980s came from developing states *outside* the ATS. Led by Malaysia, these developing states attempted to introduce the 'Question of Antarctica' for debate within the United Nations General Assembly (Tepper and Haward 2005; Hamzah and Brady 2013). This diplomatic action by developing states contested Antarctic territorial claims as being based on a 'colonialist' past, which needed to be overturned in favour of a 'common heritage of mankind' approach – essentially, a collectivist international approach to ownership of the Antarctic continent. However, the 'Question of Antarctica' issue was carefully managed during the 1980s and eventually resolved by 2004 through skilful diplomatic initiatives, both within and outside the United Nations, including those led by Australia (Haward et al. 2011).

Concurrent developments included the abandonment of the Antarctic minerals convention and its replacement by the Madid Protocol, in which Australia also played a leading role, and the decision to establish a permanent treaty secretariat. For the latter, adherence to non-militarisation played a key role in resolving the deadlock between the UK and Argentina. Historically, Argentina had objected to a treaty secretariat. In part, this was because of the risks of a supra-national body undermining its sovereign interests. Its eventual agreement to a secretariat was conditional on its location in the Argentine capital, Buenos Aires, an outcome unacceptable to the UK. The matter was eventually resolved by the UK offering to remove its objection, provided Argentina moved to a more civilian structure and operation of its Antarctic program. Thus, a potentially conflictual international issue around Antarctic territorial claims was carefully managed through oblique reference to reducing perceptions of militarisation.

States that had been antagonistic towards the Antarctic Treaty (such as Malaysia) relaxed after experiencing the benefits of accession and having their concerns about privileged access to minerals resolved by the Madrid Protocol's ban on mineral resource activities. Thus, any possibility of conflict over territorial claims or resources was averted by diplomatic engagement. Given the success to date of Article IV of the Antarctic Treaty, we believe that the risk of military conflict arising between claimant states (or the wider contracting parties to the Antarctic Treaty) is low while the Treaty remains in force. The Treaty continues until the parties decide to bring it to an end; at present, there are no indications of parties wishing to do this (Natural Environment Research Council/British Antarctic Survey 2015).

It is worth bearing in mind that only 54 of the 195 states in the world are party to the Antarctic Treaty. However, those parties represent some 80% of the world's population, including all states with current or historical connections to the continent. Parties include global powers such as the US, Russia, Japan, South Korea, France and Germany, and major developing states such as China, India, Brazil, and South Africa. Theoretically, it is possible that one or more third-party states (i.e., states outside the Antarctic Treaty) might seek to unilaterally take advantage of Antarctic resources or assert a territorial claim—perhaps over the unclaimed sector in Marie Byrd Land. However, it is likely that this would be met by considerable opposition, including from the existing claimants, other Antarctic states and, possibly, states outside the Treaty. As discussed above, the United States and the Russian Federation assert the basis of a claim to any or all of Antarctica. Even though neither country has lodged such a claim, both are likely to object to any new territorial claim asserted by a third party (just as they object to the current claims) in strong terms.

To date, none of the states outside the Antarctic Treaty have positioned themselves to make a territorial claim, or otherwise used the area below 60° South in a manner inconsistent with the Antarctic Treaty and related agreements. As noted below, the main exception has been occasional harvesting of marine living resources. Given that third-party states tend to lack a history of previous exploration or effective occupation of Antarctica, it would be difficult to see how they might have a valid claim based on international law rules on territorial acquisition. On the other hand, if an existing party to the Antarctic Treaty wished to destabilise operation of Article IV, a possible strategy might be to use a third-party state as a proxy to re-enliven conflict over Antarctic territorial claims, by supporting a third-party state to either initiate a territorial claim or to exploit resources. At present, there are no indications that parties to the Antarctic Treaty wish to re-enliven the territorial claims issue by using third parties as proxies. However, it remains a realistic possibility. If this did occur, it is moot whether such action would precipitate a military response, even under extreme provocation, given the Antarctic Treaty principles, the likely reaction of other Treaty Parties, the costs involved, and the logistical complexities.

As we noted above, unauthorised resource extraction in the ATA is a possible trigger for international conflict resulting in military intervention. Commodities that might spark such an incident include potentially extensive deposits of Antarctic minerals—including rare earth minerals, oil and gas—both on land and, more accessibly, offshore. The 29 Antarctic Treaty Consultative Parties and a further 12 non-consultative parties (a total of 41 states) have ratified the 1991 Madrid Protocol. Article 7 of the Protocol contains an indefinite prohibition on any activity relating to mineral resources, other than for scientific research. Although there is some ambiguity over the distinction between 'scientific research' and 'mineral prospecting' (which countries such as Russia have tested (Perkins and Griffin 2020), the Protocol's 'mining ban' is essentially permanent, in that it can only be removed prior to 2048 in the (very unlikely) situation of unanimous agreement of the parties.

The year 2048 is a source of endless (and often inaccurate) speculation about the future of the ATS. To be clear, Article 25(2) allows for a review conference of the Madrid Protocol to be called, if an Antarctic Treaty Consultative Party (ATCP) so requests. The earliest date that such a request might take effect is 2048, but there is no obligation to convene such a conference. Under Article 25(3), adoption of a decision to remove the mining ban (or any other substantive change to the Madrid Protocol) may only happen by a majority vote of the countries present, including three-quarters of the states which were ATCPs at the time of adoption of the Madrid Protocol in 1991. Article 25(4) provides a further barrier to change, with any decision only coming into force upon ratification, acceptance, approval, or accession of three-quarters of the ATCPs, including all states that were consultative parties at the time adoption of the Madrid Protocol in 1991. Article 25(5) also requires that, before removing the prohibition on mineral resource activities, there must be in place a binding legal regime to govern Antarctic mineral resource activities. These are very difficult voting and procedural requirements for any state wishing to remove the mining ban after 2048. While the 2048 date is often (incorrectly) suggested as a time of expiry of the Antarctic Treaty or Madrid Protocol, the more likely outcome is no significant change to the Treaty or Protocol at that time. Furthermore, even in the unlikely situation of a review conference being requested, it is foreseeable that a decision might be made to strengthen the mining ban further, such as by extending the date for a possible review, or by making the conditions under which mineral activity could proceed even harder to satisfy.

It is possible that a severe future global shortage of certain minerals will move countries within the ATS to a position where they wish to weaken or remove the current ban on mineral resource activities. In the absence of this situation, however, we suggest that significant international conflict from mining in Antarctica is more likely to come from a third-party state which is not a party to the Madrid Protocol. This would rule out many of the more obvious contenders with the requisite expertise and financial resources (i.e., Russia, United States, China, India, and Brazil), but, as discussed above, there remains the possibility of a third-party state being used as a proxy. High-value strategic commodities, such as rare earth minerals—of which China currently supplies a high percentage of the international market—may be the type of commodity to enliven the possibility of mining pressure in costly and inaccessible areas such as Antarctica, near-Earth asteroids, and the Moon. This might be the case for highly valuable rare earth mineral deposits (for instance) located on or below the seabed of the Southern Ocean. Some states argue the Madrid Protocol is ambiguous regarding whether the mining prohibition applies to offshore areas where these mineral deposits may be located (Kaye 2004). While it is unclear (due to the absence of mineral prospecting) whether Antarctica has significant quantities of such rare earth minerals, this is the type of commodity that might see a state either attempt to withdraw from the Madrid Protocol or support a third-party state to act as a proxy for prospecting and extraction activities. If unauthorised mining activities on the Antarctic continent were attempted unilaterally by third-party states, it is foreseeable that the relevant claimant state in the area could, after exhausting all other avenues, seek to intervene militarily to assert its territorial rights.

Antarctic marine living resources are also a potential cause of international tension, particularly around fishing practices. Since 1982, harvesting of finfish and krill in Antarctic waters has been governed by the ecosystem-based management and precautionary approach of the Commission for the Conservation of Antarctic Marine Living Resources (CCAMLR). The 1980 CAMLR Convention, which established CCAMLR, includes key distant water fishing states and supranational organizations, such as China, Russia, Japan, South Korea, the United States and the EU, as well as rising developing states Brazil, India, and South Africa. CCAMLR is a 'Derby' or 'Olympic fishery' (Food and Agriculture Organization of the United Nations [FAO] 2001); that is, based on a total allowable catch for a given area, rather than allocation of individual catch quotas for fishers. It operates on scientifically determined catch limits on finfish and krill based on geographic subareas of the CAMLR Convention area. The Convention area extends northwards to the Antarctic convergence, in places beyond the 60° South latitude boundary of the Antarctic Treaty. Illegal, unreported, and unregulated (IUU) fishing, particularly for high-value species such as Patagonian toothfish, was a significant issue for CCAMLR in the first decade of this century. However, the severity of the issue has reduced through effective governance actions, including catch documentation schemes to ensure the integrity of the high value toothfish catch (Haward 2004).

Notwithstanding effective management by CCAMLR, there have been instances of Australian government customs and fisheries vessels engaging in 'hot pursuit' of civilian IUU fishing vessels around Heard Island and the McDonald Islands in the Southern Ocean. These islands lie within the CCAMLR area but have exclusive economic zones that are enforceable under Australia's domestic jurisdiction. For example, in 2003, an Australian customs vessel pursued a Uruguayan fishing vessel for 3,700 nautical miles from Heard Island and the McDonald Islands to the South Atlantic. Regardless of whether such action might be characterised as military or paramilitary in nature, this was an exercise of Australian domestic legal jurisdiction against illegal fishing vessels in the area north of the ATA, and so it did not offend the Antarctic Treaty's prohibition on measures of a military nature.

The CCAMLR area is a well-governed fishery by international standards. In the short term, it is difficult to envisage how international conflict might arise out of legal or illegal fishing within the Antarctic region. It is plausible that a failure within CCAMLR to reach consensus on catch limits on key species could lead to some states fishing outside CCAMLR rules, resulting in boycotts of catch in resulting markets. But this situation would more likely remain a source of economic tension, rather than military conflict.

In the medium to long term, however, low probability/high impact events in the international system (i.e., health pandemics, extreme weather events, and famine) could place even well-governed fisheries such as CCAMLR under significant shocks. This situation could lead to more robust and militaristic enforcement action by states, particularly in the area north of the ATA. The 'cod wars' between Canada and Spain in the North Atlantic during the 1990s (Haward 2020) illustrate the point that even developed nations, with histories of otherwise good diplomatic relations, can lapse into military-style enforcement action in fisheries that are under significant stress. In particular, human-induced climate change is leading to range shifting of marine species (including high-value stocks such as tuna) to colder waters at higher latitudes (Erauskin-Extramiana et al. 2019). Range shifting of species may exacerbate international tensions already building within CCAMLR relating to fishing access, as further high-value catch moves into the Antarctic region. While we do not envisage that the CCAMLR area will become a theatre for military-style compliance action in the short term, the possibility cannot be dismissed over longer time scales.

The Southern Ocean (which includes areas extending below 60° South) is an important strategic space for the 'Antarctic gateway' states of Argentina, Australia, Chile, New Zealand, and South Africa. As noted above, in Australia's case, its southern regions are exposed to the Southern Ocean, and the Australian Government remains concerned about the long-term security of its southern maritime approaches (Press 2018). As Young (2018, 2021) points out, continuity of the non-militarisation and peaceful-use values of the Antarctic Treaty is strategically important to Australia because it lessens the need to deploy its (limited) naval and other military resources to defend the continent's southern maritime and air approaches. Similarly, the Royal Australian Navy has not assessed it necessary to procure ice-hardened ships that might operate deep into the ATA. The outcome is that most Australian defence assets can be deployed to the strategically more important northern areas of the Australian continent that intersect with South Asia and the Pacific.

While the non-militarisation and peaceful-use principles of the Antarctic Treaty appear secure for the time being, there are emerging strategic challenges in the Southern Ocean that may, in the longer-term, place pressure on the region. Brady points out that Chinese naval strategy has identified a number of alternative shipping routes that Chinese civilian supply ships might take to avoid naval 'choke points' of high strategic vulnerability (Brady 2017). Brady indicates that China has identified a shipping route that would extend into the Southern Ocean and along the area of the Tasman Sea, east of Tasmania, as an alternative to the choke point of the Malacca Straits—particularly for oil supplies transported by ship from the Middle East. If, for broader strategic reasons, the movement of civilian shipping deeper into the Southern Ocean becomes attractive for countries such as China, this could lead to increased Chinese naval presence in these areas. As Brady indicates, documents she has examined detailing this alternate shipping route suggest that it was particularly attractive as the ocean areas concerned would be under control of the Royal Australian Navy. It is interesting to consider whether Chinese authorities would find this alternative shipping route as attractive if the current political tension between Australia and China continues to build. Would a greater Chinese naval presence be forthcoming to secure Chinese shipping that operated on these routes?

The South American sector of Antarctica has been of longstanding strategic value due to the sea lanes around the Drake Passage, which straddles the boundary of the Antarctic Treaty of 60 degrees South. Historically, this has been an area of high strategic value as a shipping route between the Pacific and Atlantic Oceans. The opening of the Panama Canal in 1914 provided a more efficient route between the Pacific and Atlantic Oceans and has lessened the strategic significance of the Drake Passage. However, since 1999, when the Panama Canal came under the full sovereign control of the Panamanian Government, any political or strategic vulnerabilities in that area have potential to disrupt access. Also, US aircraft carriers are too large for Panama Canal. The location of the Drake Passage means that it will remain a strategically important area of the Southern Ocean for the countries in the immediate area (particularly Argentina and Chile); for the global shipping industry; and for Antarctic tourism which, prior to the COVID-19 pandemic, was expanding significantly.

Finally, the possibilities for use of airpower and space technologies in the area above the Antarctic continent and Southern Ocean have expanded significantly in recent years—and are likely to increase. The airspace above the ATA (at least up to the 100 km implicit legal boundary with space) is governed by the Antarctic Treaty, and so includes the provisions of Article I on peaceful use. Large modern surveillance drones, such as the MQ-4C Triton Unmanned Aircraft System, currently being acquired by the Royal Australian Air Force, have a 24-h mission capacity with a range of over 15,000 km. These unmanned aircraft systems could provide Antarctic gateway states (in particular) with significantly enhanced capabilities for maritime surveillance over the Antarctic continent and Southern Ocean. This capability can also support search and rescue operations, as well as fisheries enforcement in the

Southern Ocean. It will also give those Antarctic gateway states that are also territorial claimants (Argentina, Australia, Chile, and New Zealand) the ability to monitor Treaty compliance within their claimed territories by other states undertaking activities there. This adds to the national inspection arrangements available under the Antarctic Treaty and Madrid Protocol. It is possible that sub-Antarctic Islands might be used to base such unmanned aircraft systems, which may interest states such as France, in the Kerguelen Islands.

We recognise that modern surveillance drones have weapon-carrying variants that could be deployed to project hard military power well into the ATA. To date, many states have not had the capability to do so, and it is likely this would constitute a breach of Article I of the Antarctic Treaty. The emerging use of surveillance drones is therefore likely to create challenges to the non-militarisation of inspection regimes under the Antarctic Treaty and Madrid Protocol.

6.5 'Dual-Use' Equipment in Antarctica

As the example of the Falklands/Islas Malvinas war indicates, the Antarctic Treaty has been successful in preventing military hostilities from impinging upon the area below 60° South. We believe this is likely to continue in the short to medium term,—at least for the 54 Parties to the Antarctic Treaty.

As we have seen, Article I of the Antarctic Treaty, while clearly prohibiting any 'measures of a military nature', also allows for military personnel and equipment to operate in the ATA when used for peaceful purposes. For instance, the Argentine Defence Department is charged with Antarctic logistics and administration of most of the Argentine Antarctic research bases (Argentino 2021), and the Royal Australian Air Force (RAAF) provides logistics air support to the Wilkins ice runway near the Casey research station in East Antarctica. Australia plans to construct an all-weather runway at the Davis research base, subject to environmental approvals, which will significantly increase the support which the RAAF can provide to the Australian scientific research program, and to other programs operating within East Antarctica (McGee et al. 2021). The Royal Australian Navy also has a long history of providing logistic support to the Australian Antarctic Program.

In terms of compliance with Article I, the key factor is the *purpose* that personnel or equipment are being put to at any given time. The test here is whether the purpose is 'peaceful' or 'non-peaceful', rather than drawing inferences about the mere presence of military personnel or equipment. At different times, a military transport aircraft (such as the C-17 Globemaster operated by the RAAF) could be deployed for peaceful purposes such as search and rescue, or for non-peaceful purposes such as transporting troops into combat. The governance regime provided for in the Antarctic Treaty is *context-specific* in relation to the presence of military hardware and personnel in the ATA. This gives rise to the 'dual-use' issue, where the same personnel or equipment may be operated in a peaceful or non-peaceful fashion, depending upon context.

A similar 'dual-use' issue exists in relation to some equipment currently used at Antarctic research bases. The Antarctic continent has several significant geographic features that make it a prime site for astronomical and space research. The Antarctic ice sheet in East Antarctica rises to over 4 km, making it one of the world's highest accessible points for locating telescopes for astronomy. This high ice sheet area is also one of the most remote places on the planet, and so is one of the least affected by the ambient light and electromagnetism found closer to areas of industrial civilisation.

In recent years, concerns about militarisation of Antarctica have centred primarily on ground stations located on the Antarctic continent that support satellite technologies. Germany, Norway, the United States, Russia, India, and Japan have operated satellite ground receiving stations in Antarctica for some decades. However, concern has been raised in some quarters (Brady 2017; Bateman 2013) in relation to the construction of Chinese and Russian satellite ground receiving stations at several sites on the Antarctic continent. China and Russia have developed global positioning systems (the BeiDou and GLOSNASS networks) as alternatives to the US-controlled global positioning system (GPS) network, and Antarctic ground receiving stations might provide better coverage and accuracy for these systems. In a similar way to the GPS network, the BeiDou and GLOSNASS networks are capable of use during military activities to assist with the targeting of missiles and other weapons. This is not to suggest that Chinese and Russian ground station receiving stations have greater potential for use in this manner. It is simply to point out that, as the number of global positioning services increases, there may be greater potential for ground receiving stations in Antarctica to be used for military purposes. Further, there is also increasing concern over the militarisation of space, particularly the capacity of some states to destroy satellites with 'killer-satellites' or ground-based lasers and missiles (Dawson 2018). It is conceivable that satellite ground receiving stations in Antarctica could be used to transmit information (either up-linking or down-linking) for such space-based military purposes.

In times of military conflict, satellite ground receiving stations, which had previously been operating for peaceful purposes, could be repurposed for offensive military activity in the ways outlined above. This potential 'dual use' of satellite ground receiving stations presents a difficult problem for ensuring compliance with the non-militarisation provisions of the Antarctic Treaty. Despite recently increased efforts to undertake inspections under Article VII of the treaty, dual-use equipment is likely to remain difficult to detect—and particularly so in times of increased hostility in other regions of the world (Bergin and Press 2020). While CTBT monitoring stations clearly have a peace-keeping purpose, it would be difficult to argue that the use of satellite receiving stations to monitor military activity in times of conflict is also a peace-keeping role.

As we have noted, the Antarctic continent is an excellent venue for astronomical research and so is home to a number of telescopes, such as that located at the Kunlun Dome A Chinese base in East Antarctica. Brady raises the possibility of short-term, seamless repurposing of infrared telescopes, such as those used at Kunlun, 'to search for enemy satellites, drones and launched missiles, and to identify whether they had been shot down when targeted' (Brady 2017, p. 16), and indicates that all such telescopes located on the Antarctic continent would have a similar dual-use capacity. The short-term repurposing of telescopes for military activities would be difficult to detect without full access and analysis of the data that is captured and/or relayed. Detection of this sort would be virtually impossible with the current inspection practices followed under the Antarctic Treaty and the Madrid Protocol. Typically, these involve civilian inspectors who may not have the technical knowledge to identify potential dual-use activities, nor full access to the relevant data.

The area of the Southern Ocean below 60° South that surrounds the Antarctic continent is another potential theatre for dual-use activity. The 13th Chinese Government Five-year Plan for the Economic and Cultural Development of the People's Republic of China (2016) (People's Republic of China 2016) contains a section titled 'Safeguarding Maritime Rights and Interests' (see Part IV, pp. 118–119). This includes an aim to develop an ocean observation network for Antarctic waters. The plan indicates that, by 2021, China will:

- 'Research and develop a system for integrated deep-sea environmental monitoring and activity exploration';
- 'Complete the basic framework for a land-sea-air observation platform in the polar regions';
- 'Establish a service platform for the provision and application of information regarding the polar environments and potential polar resources'; and
- 'Work toward establishing a multidimensional global ocean observation (monitoring) system' (People's Republic of China 2016, pp. 118–119).

As Young indicates, if these plans involve the use of the system of marine observation equipment in the Southern Ocean, they could have significant strategic implications, particularly for naval and submarine forces of states adjoining the Southern Ocean, and for global naval powers such as United States (Young 2018). There is already evidence that China is deploying sophisticated marine monitoring systems on platforms in open ocean waters in highly contested areas, such as the South China Sea, which include connections to the BeiDou positioning system (People's Republic of China 2016). If a similar approach were adopted by China in the Southern Ocean, then Antarctic waters leading up to 60° South could become another site of 'dual-use' activity.

Clearly, technological developments in ocean sensors, global positioning systems, telescopes, unmanned aircraft and ships, and satellites have potential to bring new pressures on the non-militarisation of Antarctica. Potential uses of such technologies in integrated military command, control, communications, computers, intelligence, surveillance, and reconnaissance systems are evident (Johnson-Freese 2017)—and may increase as technology advances. We take up this possibility in the following two chapters, where we formulate scenarios around Antarctic militarisation.

6.6 Summary

The Southern Ocean (including maritime approaches that extend into the ATA) is of significant strategic importance for the Antarctic gateway states. These states have benefited greatly from the non-militarisation and peaceful use of Antarctica provided by the Antarctic Treaty. The Treaty has also been pivotal in protecting the interests of gateway states by ensuring that Antarctica was not used as a testing site for nuclear weapons. However, there are significant strategic vulnerabilities on the horizon for these states (in particular) and for the Southern Ocean generally. These risks include the possibility that shipping routes may be extended further south, with an accompanying naval presence. There also concerns about the construction of ocean observation systems that may have significant capabilities to identify and track shipping in the Southern Ocean and ATA, with potential military application. On the Antarctic continent itself, the threat of military activity and hostilities between states has remained low since the Antarctic Treaty came into force in 1961. Nonetheless, third-party state intervention in Antarctica remains a conceivable risk. A possible situation could involve a third-party state seeking to extract minerals, particularly high-demand rare earth metals. At present the costs and logistical challenges of attempting this remain formidable, however.

Sixty years ago, Antarctic states morphed from occasionally using Antarctica for military purposes, to only using military personnel and equipment to support peaceful activity. This quickly normalised the use of military capabilities in the region. The inspection regime was designed to verify that their use was genuinely for peaceful purposes. However, potential risks arise as global dynamics change. The more foreseeable threat to the established rules relating to peaceful use of the Antarctic continent comes from equipment and personnel that have latent dual-use capacities, which can move seamlessly from peaceful to non-peaceful activities without obvious change in the modes and practices of their day-to-day operation. This is particularly so with equipment and personnel used in the astronomical and space research area, such as ground-based satellite receiving stations and infrared telescopes.

We suggest that the potential for dual-use equipment and personnel within the ATA to inform non-peaceful activities *outside* the ATA (on land, in the oceans or in space) represents the greatest vulnerability of the ATS to a critical shift away from non-militarisation and peaceful use. Technological advances in ocean sensors, global positioning systems, telescopes, unmanned aircraft and ships, and satellites may place further pressure on the non-militarisation of the ATA, which has been largely successful for 60 years. While the inspection systems created by the Antarctic Treaty and Madrid Protocol provide some prospect of monitoring compliance with requirements regarding peaceful use (Press 2019), the modest frequency of these inspections, combined with the necessity for highly technical defence science expertise to identify instances of dual-use activity, mean that the prospects of effectively policing this activity over coming decades are slight.

In the following chapters, these findings will inform our construction of scenarios around non-militarisation and dual-use activity in Antarctica.

References

Bateman, Sam. 2013. Is Antarctica Demilitarised? *The Strategist*, 23 April. Canberra: Australian Strategic Policy Institute. https://www.aspistrategist.org.au/is-antarctica-demilitarised/. Accessed 12 August 2020.

Bergin, Anthony, and Tony Press. 2020. Eyes Wide Open: Managing the Australia-China Antarctic Relationship. *Special Report* 153 (April 2020). Canberra: Australian Strategic Policy Institute. https://www.aspi.org.au/report/eyes-wide-open-managing-australia-china-antarctic-relationship. Accessed 5 May 2021.

Blyth, J.D.M. 1952. German Raiders in the Antarctic During the War. *Polar Record* 6 (43): 399–403.

Brady, Anne-Marie. 2017. China's Expanding Antarctic Interests Implications for Australia. *The Strategist*, 17 August. Canberra: Australian Strategic Policy Institute. https://www.aspi.org.au/report/chinas-expanding-interests-antarctica. Accessed 27 July 2021.

Christie, E. W. Hunter. 1951. *The Antarctic Problem: An Historical and Political Study*. London: George Allen and Unwin.

Dahm, Michael. 2018. Chinese 'Ocean E-Stations' Deployed to the South China Sea. Asia Maritime Transparency Initiative, Center for Strategic & International Studies, 10 December. https://amti.csis.org/chinese-ocean-e-stations-deployed-south-china-sea/. Accessed 5 August 2021.

Dawson, Linda. 2018. *War in Space: The Science and Technology Behind Our Next Theater of Conflict*. Charm, Switzerland: Springer Nature.

Dodds, Klaus. 2002. *Pink Ice: Britain and the South Atlantic Empire*. London: I.B. Tauris.

Dudeney, John R., and David W. H. Walton. 2012. From Scotia to 'Operation Tabarin': Developing British Policy for Antarctica. *Polar Record* 48 (4): 342–360.

Erauskin-Extramiana, Maite, Haritz Arrizabalaga, Alistair J. Hobday, Anna Cabré, Leire Ibaibarriaga, Igor Arregui, Hilario Murua, et al. 2019. Large-scale Distribution of Tuna Species in a Warming Ocean. *Global Change Biology* 25: 2043–2060.

Food and Agriculture Organization of the United Nations [FAO]. 2001. Fisheries Management and Safety. *Safety at Sea as an Integral Part of Fisheries Management*, FAO Fisheries Circular No. 966, Ch. IV. Rome: FAO. http://www.fao.org/3/x9656e/X9656E03.htm. Accessed 5 May 2021.

Gobierno Argentino. 2021. Bases Antárticas argentinas. Ministerio de Relaciones Exteriores, Comercio Internacional y Culto. https://www.cancilleria.gob.ar/es/iniciativas/dna/antartida-argentina/bases. Accessed 5 May 2021.

Haddelsey, Stephen, with Alan Carrol. 2014. *Operation Tabarin: Britain's Secret Wartime Expedition to Antarctica, 1944–46*. Stroud, UK: The History Press.

Hall, H. Robert. 1994. 'International Regime Formation and Leadership: The Origins of the Antarctic Treaty', Ph.D. thesis, University of Tasmania. https://eprints.utas.edu.au/11401/. Accessed 5 August 2021.

Hamzah, B.A. 2013. The Malaysian Journey to the Antarctic: A Glimpse at Public Policy Dynamics. In *The Emerging Politics of Antarctica*, ed. Anne-Marie. Brady, 96–109. Abingdon, UK: Routledge.

Haward, Marcus. 2004. IUU Fishing: Contemporary Practice. In *Oceans Management in the 21st Century: Institutional Frameworks and Responses*, eds. Alex G. Oude Elferink and Donald R. Rothwell, 87–106. Leiden: Martinus Nijhoff Publishers.

Haward, Marcus. 2020. *Governing Oceans in a Time of Change: Fishing for the Future?* Cheltenham: Edward Elgar.

Haward, Marcus, and David Mason. 2011. Australia, the United Nations and the Question of Antarctica. In *Australia and the Antarctic Treaty System: Fifty Years of Influence*, ed. Marcus Haward and Tom Griffiths, 202–221. Sydney: UNSW Press.

Johnson-Freese, Joan. 2017. *Space Warfare in the 21st Century: Arming the Heavens*. New York: Routledge.

Kaye, Stuart. 2004. Implementing High Seas Biodiversity Conservation: Global Geopolitical Considerations. *Marine Policy* 28 (3): 221–226.

King, Matt 2021. Personal Communication. 1 November.

Lüdecke, Cornelia. 2012. Investigation of the Unknown: The Flight Programme of the German *Schwabenland* Expedition 1938/39. *The Polar Journal* 2 (2): 231–333.

McGee, Jeffrey, Marcus Haward, and Anthony Bergin. 2021. Gamechanger: Australian Leadership for All-Season Air Access to Antarctica. *Strategic Insights* 160 (April 2012). Canberra: Australian Strategic Policy Institute. https://www.aspi.org.au/report/gamechanger-australian-leadership-all-season-air-access-antarctica. Accessed 5 May 2021.

Natural Environment Research Council/British Antarctic Survey. 2015. The Antarctic Treaty. https://www.bas.ac.uk/about/antarctica/the-antarctic-treaty. Accessed 27 July 2021.

People's Republic of China. 2016. The 13th Five-year Plan for Economic and Social Development of the People's Republic of China (2016–2020). Trans. Compilation and Translation Bureau, Central Committee of the Communist Party of China, Beijing. https://policy.asiapacificenergy.org/node/2509. Accessed 28 July 2021.

Perkins, Robert, and Rosemary Griffin. 2020. Russia Stokes Political Tensions with Hunt for Antarctic Oil, 5 May. *S&P Global Platts*. https://www.spglobal.com/platts/en/market-insights/latest-news/oil/022120-russia-stokes-political-tensions-with-hunt-for-antarctic-oil. Accessed 27 July 2021.

Press, A. J. 2018. Australia's Most Southern Shores: The Strategic Geography of Australia and the Southern Ocean. In *Australian Contributions to Strategic and Military Geography*, eds. Stuart Pearson, Jane Louise Holloway, Richard Malcolm Thackway, 129–142. Charm, Switzerland: Springer.

Press, Tony. 2019. Australia Wants to Install Military Technology in Antarctica - Here's Why That's Allowed. *The Conversation*, 23 August. Melbourne: The Conversation Media Group. https://theconversation.com/australia-wants-to-install-military-technology-in-antarctica-heres-why-thats-allowed-122122 . Accessed 12 August 2020.

Rack, Ursala. 2015. Polar Expeditions. In *Exploring the Last Continent: An Introduction to Antarctica*, ed. Daniela Ligget, Bryan Storey, Yvonne Cook, and Veronika Meduna, 305–325. Charm, Switzerland: Springer.

Royal Institute of International Affairs. 1953. *Chronology of International Events and Documents* 9(5): 136–137.

Tepper, Rohan, and Marcus Haward. 2005. The Development of Malaysia's Position on Antarctica: 1982–2004. *Polar Record* 41 (2): 113–124.

Young, Claire. 2018. What's China Up to in Antarctica? *The Strategist,* 20 September. Canberra: Australian Strategic Policy Institute. https://www.aspistrategist.org.au/whats-china-up-to-in-antarctica/. Accessed 12 August 2020.

Young, Claire. 2021. *Eyes on the Prize; Australia, China and the Antarctic Treaty System.* Policy Briefs, 16 February. Sydney: Lowy Institute.

Chapter 7
Antarctic Militarisation: Scenario Analysis

Abstract In this chapter, we illustrate a process for constructing scenarios on Antarctic militarisation in the classical geopolitical tradition. We augment the standard intuitive logics '2 × 2' process in several ways to address some of the difficulties encountered when applying this technique to Antarctic geopolitics. We identify around 50 driving forces relevant to Antarctic militarisation, grouped across nine themes. Of these, two driving forces stand out as the most uncertain and important—these are our top two 'critical uncertainties'. Our first critical uncertainty is the level of stability of the international system, relevant to the ATS. Here, a decrease in the stability of the international system may hasten a shift to multipolarity and great power competition, particularly between China, Russia, and the United States—with potential for spill over effects for the ATA. Our second critical uncertainty is the level of strategic advantage from Antarctic militarisation. This concerns the level of military/strategic advantage that states may gain from using Antarctica and the Southern Ocean for non-peaceful activities such as military research and substantive military activity. Using our novel 'IL 5' approach, we take these two critical uncertainties (at their high, medium, and low values) to construct five scenarios on Antarctic militarisation.

Keywords Militarisation of Antarctic Treaty Area · Driving forces of Antarctic militarisation · Command, control, communications and reconnaissance (CCCR) systems in Antarctica · Predetermined elements and critical uncertainties · Scenario logic for Antarctic militarisation

7.1 Introduction

In Chap. 6, we provided background on the issue of militarisation in Antarctica and the Southern Ocean. We presented an historical overview of the (very limited) military activity in the Antarctic Treaty Area (ATA) up until the 1950s, and the diplomatic moves to form the Antarctic Treaty 1959, which created a very significant non-militarised zone in the southern hemisphere. We set out some current issues and potential future pressures.

J. McGee et al., *The Future of Antarctica*, Springer Polar Sciences,
https://doi.org/10.1007/978-981-16-7095-4_7

Here, we draw on this material and the scenario methodologies outlined in Chaps. 3–5 to illustrate a systematic approach to constructing scenarios in the classical geopolitical tradition. We acknowledge there are limitations to this exercise that would be overcome in a 'real world' scenario process. First, in the real world, we would undertake a more comprehensive exercise to identify and prioritise the key driving forces of Antarctic geopolitics. We sketched some of the salient issues in Chap. 5, then settled on militarisation as an important but manageable sub-topic to illustrate the scenario process. Many other issues could have been selected. Second, in the real world, we would engage with a diverse set of participants[1] and develop these scenarios with them, through multiple iterations. In this book, we have sought feedback from a small number of specialists with expertise relevant to the militarisation issue. Noting these limitations, we hope to illustrate a pragmatic and policy-relevant approach for stakeholders interested in this topic.

In Chap. 4, we outlined the nine basic steps of the 'IL 2 × 2' technique:

1. Identify your scenario standpoint;
2. Set your time horizon;
3. Identify the key driving forces;
4. Distinguish critical uncertainties from predetermined elements;
5. Rank critical uncertainties by importance and uncertainty;
6. Select your general approach/es;
7. Construct the 'IL 5' scenario matrix;
8. Develop the scenario narratives; and
9. Wind tunnel test current strategy and prepare policy advice.

The following sections of this chapter will apply Steps 1–8 to illustrate how a scenario analysis for Antarctic militarisation might be undertaken. Then, in Chap. 8, we will unpack the scenarios in detail and consider some policy advice that flows from them.

7.2 Step 1: Identify Your Scenario Standpoint

As noted in Chap. 4, effective scenarios (of the type illustrated here) are conducted from one perspective. The perspective we select may be that of a single actor, or a coherent coalition of interests. Our perspective draws on the disciplines of international law, geopolitics, and governance. Importantly, although we are Australian, we do not represent the official views of the Australian Government, which are instead reflected in its official policy documents.

The Australian Government's current policy in this area was articulated in the *2016 Australian Antarctic* Strategy and *20-year Action Plan* (Commonwealth of

---

[1] While a diverse set of scenario participants is more likely to challenge common assumptions and avoid 'groupthink', the extent of participation by external stakeholders and experts can be subject to restrictions on access to information and confidentiality requirements. To overcome this, organisations may wish to consider conducting *both* restricted and less restricted scenario sessions.

Australia 2016a). As far as we are aware, this is one of the few instances where states have articulated their national interests in Antarctica comprehensively. Within that strategy, one of the key policy objectives is to 'Maintain Antarctica's freedom from strategic and/or political confrontation' (Commonwealth of Australia 2016a, p. 17). Young observes that a core strategic interest for Australia is for the ATA to remain non-militarised (Young 2021). As noted in Chap. 6, this allows for Australia's limited defence assets to be focused on the strategically important eastern, northern, and western sea approaches of the Pacific and Indian Oceans (Medcalf 2020; Bergin and Press 2020). It also avoids the need for Australia to maintain a 'three-ocean navy'. A non-militarised Antarctica furthers another of Australia's key policy objectives under its strategy, which is to 'Support a strong and effective Antarctic Treaty system' (Commonwealth of Australia 2016a, p. 17).

On the issue of Antarctic militarisation, the scenario standpoint we adopt is that of states whose policy and actions preserve and enhance the current rules and norms of the Antarctic Treaty System (ATS). This includes promoting the peaceful use of Antarctica and the Southern Ocean. We hope this scenario standpoint is also relevant to other stakeholders—but we acknowledge that what constitutes a critical uncertainty for us may not hold true for others. Nonetheless, we hope that these scenarios will strike a chord with the perspectives of other states in the ATS—especially the states geographically proximate to the Antarctic continent (New Zealand, Chile, Argentina, and South Africa) and those states with sub-Antarctic islands, such as France, Norway, and the United Kingdom. Although this analysis is consciously limited to a state-centric perspective, we acknowledge that other voices are important. We also hope there is some value here for sub-state actors (such as civil society groups) with interest in the future of Antarctica and the Southern Ocean.

7.3 Step 2: Set Your Time Horizon

The time horizon for these scenarios is to roughly a decade. The convergence of uncertainty, complexity and the passage of time means this is far enough into the future to take us beyond the limits of standard prediction, but not so far ahead as to be untethered from current developments in geopolitics and military technology.

7.4 Step 3: Identify Driving Forces of Antarctic Militarisation

The third step is to identify the key driving forces likely to influence Antarctic militarisation over the target period. In Chap. 6, we used the term 'militarisation' in Antarctica to refer to those activities that involve the use of military personnel and/or equipment for non-peaceful purposes. These activities might occur on land, ice-covered areas, oceans and in airspace of the ATA. The outcome of this non-peaceful activity might be realised directly within the ATA, or indirectly in other parts of the world.

As we saw in Chap. 6, because of its geographic remoteness and harsh climate, Antarctica has only a modest history of military activity. Perhaps more importantly, the *potential* for military hostilities in Antarctica has played a prominent role in the minds of policy makers; in particular, Cold War competition between the Soviet Union and the United States was a key driver for development of the 1959 Antarctic Treaty.

Since 1961, when the Antarctic Treaty came into effect, overt military activity has been absent from the ATA. As we saw in Chap. 6, this is consistent with the prohibition on activities of a military nature contained in Article I of the Treaty. However, it is important to note that, unlike the situation in other arms control agreements, for most of the life of the Treaty there has been limited (and perhaps declining) incentive for states to carry out military activities within the ATA. Indeed, military activity of strategic importance has been carried out more effectively from places outside the ATA (Parsons 1987). Thus, for the last 60 years, the prohibition on military activity within the ATA has aligned with and suited the strategic interests of key Antarctic states.

We suggest that pressures for militarisation in Antarctica over the next decade may differ significantly from those of the last 60 years, however. As we saw in Chap. 6, with increasing instability in the international system (including the return of great power competition) and emerging military technologies, some states may reassess whether there are strategic advantages to carrying out some types of military activity below 60° South. In particular, activities that contribute to wider military command, control, communications and reconnaissance (CCCR) networks may become increasingly attractive. A key point for our scenarios is that most of the driving forces for Antarctic militarisation are likely to come from critical uncertainties around geopolitics and military technology that come from *outside* Antarctica and the ATA, rather than from within.

As we have seen, scenario analysis requires us to identify these driving forces (and understand their dynamics) by gathering data and assessing it using expert judgement. To establish our initial evidence base, and to reduce confirmation bias among the analysts, it is important to scan for any available quantitative data, to identify any broad trends and patterns (Tetlock and Gardner 2015). Given political and security sensitivities in relation to Antarctic militarisation, however, only limited data is in the public domain. Indeed, publicly available data on Antarctic military activities is limited to the annual reporting of states to the Antarctic Treaty on their intended use of military personnel in support of their (peaceful) national Antarctic research programs. As discussed in Chap. 6, military personnel and equipment have long been used to support Antarctic science and this is expressly provided for under the Antarctic Treaty. However, it is possible that military personnel who are ostensibly supporting scientific research in Antarctica could also carry out military-related tasks on a covert basis. We are not aware of publicly available, contemporary evidence in relation to this. Although quantitative data is available from the Antarctic Treaty Secretariat website on reported military personnel used in national Antarctic programs (and this could be plotted across time), we believe it would have limited probative value for the construction of these scenarios, so has not been relied on here. Data limitations of this type are not fatal for scenarios, which become useful in the zone *beyond* standard prediction. What matters is that the scenario narratives are *plausible* outcomes, given the evidence base that is available.

Qualitative data is also used to inform the judgment of the scenario analyst. The available qualitative data with most probative value on military activities in the ATA is drawn from existing open-access government reports (Layton et al. 2019; Commonwealth of Australia 2006, 2016b, 2018, 2020a, b; Hodgman 2016; Bieniek et al. 2020) and expert commentaries (Young 2018, 2021; Bergin and Press 2019; Bateman 2013; Brady 2017, 2018; Press 2020; Bergin et al. 2019). This was supplemented by broader political reports, analysis and commentary on global and regional geopolitical trends, patterns, and emerging issues (Medcalf 2020; Luce 2017; Allison 2017; Mullerson 2017; Mearsheimer 2018, 2019; Brands and Edel 2019; Taylor 2018; Haass 2020; Buchanan and Burke 2021). These are particularly useful for situating Antarctic militarisation within broader, but related, trends and patterns occurring within the international system.

After considering the available data, we identified the following potential driving forces for militarisation in Antarctica. These are ranked (in roughly descending order of likelihood) using our subjective assessment after consultation with a small number of subject matter experts. Again, this ranking is for illustrative purposes, and is not intended to replace a more rigorous (and iterative) prioritisation of drivers based on well-developed criteria, qualitative and quantitative data, analysis, stakeholder consultation, and expert judgement. These drivers are also grouped under the nine broad categories of driving forces outlined in Chaps. 4 and 5: political, economic, social, technological, environmental, legal, geographic, strategic and historical.

Finally, taking up our observations (from Chap. 5) about the different ways that a driving force can operate within a scenario, we will also categorise each driving force here according to its tendency to impel **(I)** or constrain **(C)** militarisation in Antarctica. Unless otherwise stated, the principal actors in these driving forces are nation states with a presence in Antarctica and the Southern Ocean (Table 7.1).

Having identified our driving forces, we will now turn to the next step in the scenario construction process.

7.5 Step 4: Classify Driving Forces as Predetermined Elements or Critical Uncertainties

The next step involves categorising the driving forces listed above, according to whether they are *predetermined elements* or *critical uncertainties*. To recap, predetermined elements are driving forces that might reasonably be assumed to be relatively stable and easy to predict within the temporal scale of the target period (Schwartz

Table 7.1 Driving forces for militarisation in Antarctica

(1)	**Political** driving forces	
Stability of the international system, relevant to the ATS	Stability of the international system may decrease, with a shift to multipolarity and great power competition, particularly between China, Russia, and the United States, with potential for spill-over effects into the ATA. (A range of related political driving forces contributes to this, such as the foreign policy objectives of powerful states, and how these objectives are perceived by other states, etc.)	**I**
State control over international spaces	Interests of key states (for example, China, Japan, Russia, South Korea, and the United States) in exercising greater governance control over international spaces, to keep open the possibility of resource extraction	**I**
China's desire for oceans influence	Desire of China to demonstrate international leadership in oceanography and ocean observation systems	**I**
State and NGO desire for increased monitoring	Political pressure from some states and international NGOs for more frequent inspections of ships, aircraft, and research stations under Article VII of the Antarctic Treaty and Article 14 of the Madrid Protocol	**C**
(2)	**Economic** driving forces	
Developing states' demand for marine living resources	High rates of economic growth in large, populous developing states are increasing demand for protein, including from high-value fisheries (such as Southern Ocean toothfish and icefish) and krill, either as animal feedstock or human food	**I**
State interest in extraction of Antarctic mineral resources	Interest in mineral resource extraction by some Antarctic Treaty parties, and possible use of third-party states as a proxy, to destabilise the prohibition on mineral resource activities. A range of economic driving forces contributes to this, such as global, regional, and state-level supply and demand for minerals for energy and manufacturing, and innovation in extraction technologies for polar regions	**I**

(continued)

Table 7.1 (continued)

Global demand for rare earth elements (REEs)	REEs provide essential components for many technologies, including magnets, batteries, lasers, electronics, and aerospace alloys (Huleatt 2019). Demand is rising sharply with increased production of mobile phones, electric cars, and renewable energy (wind turbines and photovoltaics). Concentration in the supply of REEs within a small number of states (principally China) could make mineral prospecting in Antarctic more attractive for states wishing to diversify their supply options. Antarctic offshore minerals are the most accessible target for extraction	**I**
(3)	**Social** driving forces	
Level of domestic nationalist ambitions within China and Russia	Domestic nationalist desires for a return to global leadership and international respect. Examples include China's ambition to overcome the 19-20th 'century of humiliation' at the hands of Western powers (Allison 2017), and Russian desire for a return to great power status in the international system following collapse of the Soviet Union	**I**
Developing countries' demand for fish	Rising affluence of middle-class populations in large, developing states could create pressure on those states to expand their distant water fishing operations to meet increasing demand for protein	**I**
Developed countries' demand for REEs	Continued expansion in consumption levels in developed states is increasing demand for high technology products manufactured from REEs	**I**
(4)	**Technological** driving forces	
Use of ground receiving stations for military purposes	Developments in global positioning systems lead to use of ground receiving stations in Antarctica and the Southern Ocean to support military command, control, communications, and reconnaissance (CCCR) platforms	**I**
Use of links to military satellites from ground stations	Developments in military satellite technologies lead to use of ground receiving stations in Antarctica for down-link/up-link of military communications	**I**

(continued)

Table 7.1 (continued)

Use of infrared telescopes for military purposes	Developments in infrared telescopes lead to their use in target tracking and military surveillance applications	**I**
Use of ocean observation systems for military surveillance	Developments in ocean observations systems lead to use of these technologies within the Southern Ocean and ATA for military maritime surveillance	**I + C**
Use of unmanned aircraft systems for military surveillance and offensive capability	Developments in unmanned aircraft systems enable greater projection of aerial surveillance deep into the ATA, as well as capability to deliver ordinance payloads. This leads to their increasing use in Antarctica through integration in military CCCR systems	**I**
Capability to detect traditional military activity	Increasing surveillance capabilities mean that traditional military activity in Antarctica and the Southern Ocean (such as the presence of weapons or personnel undertaking combat-oriented activity) is easier to detect. This is a constraint on some forms of military activity	**C**
Range and endurance of vessels, aircraft, and weapon systems	Replacement of existing capabilities with vessels, aircraft, and weapon systems with greater range and/or endurance provides further operational reach and extended operating seasons in Antarctica and the Southern Ocean	**I**
Presence and capabilities of icebreakers	Release of nuclear-powered icebreakers from Arctic fleets provides year-round access to the Southern Ocean and Antarctic coastline, with greater capacity leading to increased rivalry for maritime influence in Antarctica	**I**
(5)	**Environmental** Driving Forces	
Severity of climate change impacts	The magnitude and speed of increases in global mean temperatures from human activity (principally greenhouse gas emissions) remains uncertain. Projections for coming decades remain within a wide range, due to the large number of variables. The potential distribution and severity of impacts varies accordingly. Political, economic, and strategic changes may result (see below)	**I**

(continued)

Table 7.1 (continued)

Impact of climate change on demand for REEs	Climate change drives demand for electric vehicles and renewable energy technologies (wind turbines and photovoltaics). This increases demand for REEs used in the production of magnets and batteries. Relates to mining of REE deposits in Antarctica (see **Economic** driving forces, above)	**I**
Impact of climate change on access to Antarctica	Climate change makes some terrestrial and marine areas of Antarctica more accessible for human activities for greater parts of the year	**I**
Impact of climate change as a threat multiplier for food security	Climate change acts as a threat multiplier to food security, placing increased pressure on marine harvesting in the Southern Ocean	**I**
Impact of climate change on species range shift	Climate change causes range shifting of key marine species and fishing activities into higher latitudes of the Southern Ocean	**I**
Level of tension caused by geoengineering in Antarctica	Unilateral or mini-lateral moves by states to undertake geoengineering activities within the ATA will increase international tension if such action is opposed by other states. Geoengineering includes: ocean iron fertilisation (to facilitate carbon dioxide removal from atmosphere and storage on ocean floor); marine cloud brightening (solar radiation management measures to slow sea ice loss); and ice sheet stabilisation proposals (to slow the impacts of climate change on Antarctic ice sheets)	**I + C**
Incidence of IUU fishing	Increased levels of illegal, unreported, and unregulated (IUU) fishing place stress on Southern Ocean finfish stocks	**I**
Level of demand for Antarctic MPAs	International demand for climate resilience increases demand for Antarctic Marine Protected Areas (MPAs)	**I + C**

(continued)

Table 7.1 (continued)

(6)	**Legal** driving forces	
Adherence by parties to treaty obligations relating to Antarctica	Treaty obligations in force under the ATS constrain parties from militarising the ATA. In addition, some treaty provisions diffuse key issues with potential to trigger militarisation. In particular: Article I of the Antarctic Treaty bans military activity for non-peaceful purposes; Article IV provides a mechanism for respecting various positions on territorial sovereignty; and Article 7 of the Madrid Protocol imposes a ban on mining within the ATA	**C**
Adequacy of the definition of 'measures of a military nature'	Innovations in military technologies fail to be captured adequately within the definition of 'measures of a military nature' in the Antarctic Treaty. This is due to lack of development in this definition or changing interpretations of it by some states	**I**
Capacity of the inspection regime to detect some types of breach of Art IV	Military activities are broader in scope than when the Antarctic Treaty was drafted in the late 1950s (see **Technological** driving forces, above). Some contemporary military activity is almost invisible to external scrutiny, so some breaches of Article IV remain undetected, or are difficult to prove	**I**
Use of civilians for military purposes	Some civilian research and operation of ground receiving stations may amount to a non-peaceful use, and thus constitute measures of a military nature. Lack of clarity in this area makes detection of breaches difficult	**I**
Compliance with information exchange	Poor compliance (by some Parties) with information exchange obligations weakens the inspection / verification system	**I**
Capability of inspection regime to detect dual use of equipment	Inspection and monitoring practices under the Antarctic Treaty and Madrid Protocol are infrequent and limited in scope. Inspection teams lack sufficient access and technical expertise to identify whether certain types of equipment are being operated for military purposes	**I**

(continued)

Table 7.1 (continued)

Extent to which the current standard for 'peaceful use' is maintained in international law	With continuing militarisation of the near space domain (satellites in near space orbits used in military communications and anti-satellite systems), the norm of 'peaceful use' of space in international space law is re-interpreted to mean 'non-aggression', rather than the higher standard of 'non-militarisation.'	**I**
Level of contestation over the purpose of CCAMLR	Contestation by key countries (China, Japan, Norway, and Russia) over the object and purpose of Antarctic marine living resources. For example, disagreement over the definition of 'conservation' in Art II(1) of the CAMLR Convention	**I**
Desire of Parties to weaken the current mining ban at a future Review Conference, or by leaving the Madrid Protocol	The possibility of a Review Conference for the Madrid Protocol (available from 2048 onwards) is viewed by some countries as an opportunity to weaken or remove the current prohibition on mineral resource activities. A country might also leave the Madrid Protocol in order to remove legal obstacles to it engaging in mineral resource activities	**I**
(7)	**Geographic** driving forces	
Impacts of isolation and rough seas	Antarctica's isolated location, surrounded by large expanses of frequently rough Southern Ocean, add to the costs and logistical barriers of military activity on the continent—particularly in East Antarctica	**C**
Impacts of extreme climate	Antarctica's extreme climate, with very low temperatures, frequent high winds, and long periods of darkness over winter, makes military activity costly, logistically complex, and hazardous	**C**
Extent of ice on land and at sea	The presence of thick ice sheets, glaciers, and ravines on 98% of the continent's surface, together with sea ice and icebergs surrounding the continent's shores, make maritime access to the continent and movement across it logistically challenging, requiring costly specialist equipment such as icebreaking ships	**C**

(continued)

Table 7.1 (continued)

(8)	**Strategic** driving forces	
Strategic advantage from Antarctic militarisation	Level of military/strategic advantage from use of Antarctica and the Southern Ocean for non-peaceful activities such as military research and substantive military activity	**I**
Extent of strategic advantage from militarisation of satellites	Extent of military/strategic advantage from further militarisation of near space satellite orbits (connected to Antarctic ground stations)	**I**
Degree of reliance on the Southern Ocean as an alternative to maritime choke points	Potential greater reliance on Southern Ocean shipping routes by military and civilian vessels, to avoid maritime choke points, such as the Suez Canal, Panama Canal, Drake Passage, and the Malacca Straits	**I**
Adequacy of monitoring for nuclear explosions	Antarctica's 12 monitoring facilities to detect nuclear explosions (set up under the 1996 Comprehensive Nuclear Test Ban Treaty—not yet in force) continue to act as an additional constraint on nuclear weapons testing within the ATA	**C**
Level of strategic advantage for 'gateway' states from non-militarisation	For Antarctic 'gateway' states (Argentina, Australia, Chile, New Zealand, and South Africa), non-militarisation of the ATA and Southern Ocean continues as a strategic priority, as it reduces threats from southern approaches, and enables military resources to be focused toward higher-risk areas	**C**
Intensity of competition over sub-Antarctic islands	Sovereignty tensions over the sub-Antarctic islands lead to more extensive occupation, or other forms of competition over these territories	**I**

(continued)

Table 7.1 (continued)

Level of strategic value of Antarctic assets	The lack of strategic value of most National Antarctic Program assets in Antarctica—such as research stations and depots—acts as a disincentive (constraint) on military activity	
(9)	**Historical** driving forces (with ongoing influence)	
Intensity of geopolitical tension caused by overlapping territorial claims	Geopolitical tensions from the overlapping territorial claims of Argentina, Chile, and the United Kingdom	**I**
Extent of differences on the status of territorial claims	Longstanding, entrenched and continuing differences in states' stances over the status of territorial sovereignty in Antarctica	**I**
Level of support for Antarctica as common heritage—for *exploitation*	The view of some states that Antarctica is 'common heritage', that its resources (including marine resources) are global property, and that these resources should be made available for exploitation and distribution (from the 1980s onwards)	**I**
Level of support for Antarctica as common heritage—for *preservation*	The view of some states that Antarctica is 'common heritage', that its environmental and aesthetic values are a global asset, and that these values should be preserved forever (from the 1980s onwards)	**C + I**

1996). In contrast, critical uncertainties (Schwartz 1996; van der Heijden 1996) are driving forces that are relatively unpredictable and inherently incapable of being estimated or measured accurately within the target period—even if improved information and data might become available. Critical uncertainties are important variables in our scenarios. If we can identify the most important critical uncertainties in a field—and anticipate what plausible impacts those uncertainties may have on the future—we can develop appropriate strategies and policy responses.

In the context of Antarctic geopolitics, our significant critical uncertainties are also likely to have a decisive impact upon conflict/cooperation outcomes. In addition, to qualify as a critical uncertainty, it must be impossible for a key actor to manage or influence that uncertainty alone. In Chap. 4, we saw that critical uncertainties are selected for use in the IL 2 × 2 because they can be defined clearly; they can occur (plausibly) at very low and very high values; and they are causally independent of each other—otherwise, there is too much complexity within the logic of the model. Also in Chap. 4, we suggested that a simple binary distinction between a predetermined element and a critical uncertainty may be too crude. Why? Because our level of uncertainty about a driving force may vary, according to the degree of its *complexity*, or according to the *adequacy of information* we have about it. For this reason, we suggest that most driving forces sit somewhere along an uncertainty spectrum. We will illustrate this below.

7.6 Step 5: Rank Driving Forces by Importance and Uncertainty

As we saw in Chap. 4, in this step, we need to identify which of our critical uncertainties exhibit the greatest potential to shape our scenario outcomes. To do this, we will select the two uncertainties that we assess as the most *uncertain* and the most *important* for use in our IL 2 × 2 matrix. We illustrate our process in Table 7.2. Using subjective judgement from a limited set of experts, we have scored each driving force on a four-point Likert-type scale.[2]

In Column A, we score *uncertainty* as follows:

- 1 = very certain
- 2 = fairly certain
- 3 = fairly uncertain
- 4 = very uncertain.

[2] Unlike a five-point Likert-type scale, a four-point scale forces us to decide whether the driving force is *predominantly* certain or uncertain; and *predominantly* important or unimportant. This decision must be made if we are to use the IL 2 × 2 technique. A four-point scale still allows for a more nuanced analysis of the potential behaviour of driving forces within our scenarios than a simple binary choice between these categories.

Table 7.2 Scoring the driving forces for militarisation in Antarctica

(1)	**Political** driving forces					
		(A)		(B)	(C)	(D)
Stability of the international system, relevant to the ATS	Very uncertain	4	Very important	4	**CU**	**16**
State control over international spaces	Fairly uncertain	3	Fairly important	3	**CU**	**9**
China's desire for oceans influence	Very certain	1	Fairly important	3	*PE*	*3*
State and NGO desire for increased monitoring	Fairly certain	2	Fairly important	3	*PE*	*6*
(2)	**Economic** driving forces					
Developing states' demand for marine resources	Fairly certain	2	Fairly important	3	*PE*	*6*
State interest in extraction of Antarctic mineral resources	Fairly uncertain	3	Fairly important	3	**CU**	**9**
Global demand for rare earth elements (REEs)	Fairly certain	2	Fairly important	3	*PE*	*6*
(3)	**Social** driving forces					
Level of domestic nationalist ambitions within China and Russia	Fairly certain	2	Fairly important	3	*PE*	*6*
Developing countries' demand for fish	Fairly certain	2	Fairly important	3	*PE*	*6*
Developed countries' demand for REEs	Fairly certain	2	Fairly important	3	*PE*	*6*
(4)	**Technological** driving forces					
Use of ground receiving stations for military purposes	Fairly uncertain	3	Very important	4	**CU**	**12**
Use of links to military satellites from ground stations	Fairly uncertain	3	Very important	4	**CU**	**12**
Use of infrared telescopes for military purposes	Fairly uncertain	3	Very important	4	**CU**	**12**

(continued)

Table 7.2 (continued)

Use of ocean observation systems for military surveillance	Fairly certain	2	Fairly important	3	*PE*	*6*
Use of unmanned aircraft systems for military surveillance and offensive capability	Fairly uncertain	3	Very important	4	**CU**	**12**
Capability to detect traditional military activity	Fairly certain	2	Fairly important	3	*PE*	*6*
Range and endurance of vessels, aircraft, and weapon systems	Fairly certain	2	Fairly important	3	*PE*	*6*
Presence and capabilities of icebreakers	Fairly uncertain	3	Fairly important	3	**CU**	**9**
(5)	**Environmental** driving forces					
Severity of climate change impacts	Fairly uncertain	3	Very important	4	**CU**	**12**
Impact of climate change on demand for REEs	Fairly uncertain	3	Fairly important	3	**CU**	**9**
Impact of climate change on access to Antarctica	Fairly uncertain	3	Fairly important	3	**CU**	**9**
Impact of climate change as a threat multiplier for food security	Fairly certain	2	Fairly important	3	*PE*	*6*
Impact of climate change on species range shift	Fairly certain	2	Fairly unimportant	2	*PE*	*4*
Level of tension caused by geoengineering in Antarctica	Fairly uncertain	3	Fairly unimportant	2	**CU**	**6**
Incidence of IUU fishing	Fairly uncertain	3	Fairly important	3	**CU**	**9**
Level of demand for Antarctic MPAs	Fairly uncertain	3	Fairly important	3	**CU**	**9**

(continued)

Table 7.2 (continued)

(6)	**Legal** driving forces					
Adherence by parties to treaty obligations relating to Antarctica	Fairly uncertain	3	Very important	4	**CU**	**12**
Adequacy of the definition of 'measures of a military nature'	Fairly uncertain	3	Fairly important	3	**CU**	**9**
Capacity of the inspection regime to detect some types of breach of Art IV	Fairly uncertain	3	Very important	4	**CU**	**12**
Use of civilians for military purposes	Fairly uncertain	3	Very important	4	**CU**	**12**
Compliance with information exchange	Fairly uncertain	3	Fairly important	3	**CU**	**9**
Capability of inspection regime to detect dual use of equipment	Fairly uncertain	3	Very important	4	**CU**	**12**
Extent to which the current standard for 'peaceful use' is maintained in international law	Fairly uncertain	3	Fairly important	3	**CU**	**9**
Level of contestation over the purpose of CCAMLR	Fairly certain	2	Fairly important	3	*PE*	*6*
Desire of Parties to weaken the current mining ban at a future Review Conference	Fairly uncertain	3	Fairly important	3	**CU**	**9**
(7)	**Geographic** driving forces					
Impacts of isolation and rough seas	Very certain	1	Very important	4	*PE*	*4*
Impacts of extreme climate	Very certain	1	Very important	4	*PE*	*4*
Extent of ice on land and sea	Fairly uncertain	3	Very important	4	**CU**	**12**
(8)	**Strategic** driving forces					
Strategic advantage from Antarctic militarisation	Very uncertain	4	Very important	4	**CU**	**16**

(continued)

Table 7.2 (continued)

Extent of strategic advantage from militarisation of satellites	Fairly certain	2	Very important	4	*PE*	*8*
Degree of reliance on the Southern Ocean as an alternative to maritime choke points	Very uncertain	4	Fairly unimportant	2	**CU**	**8**
Adequacy of monitoring for nuclear explosions	Fairly certain	2	Very important	4	*PE*	*8*
Level of strategic advantage for 'gateway' states from non-militarisation	Very certain	1	Very important	4	*PE*	*4*
Intensity of competition over sub-Antarctic islands	Fairly uncertain	3	Fairly important	3	**CU**	**9**
Level of strategic value of Antarctic assets	Fairly certain	2	Fairly important	3	*PE*	*6*
(9)	**Historical** driving forces (with ongoing influence)					
Intensity of geopolitical tension caused by overlapping territorial claims	Fairly uncertain	3	Fairly important	3	**CU**	**9**
Extent of differences on the status of territorial claims	Very certain	1	Fairly important	3	*PE*	*3*
Level of support for Antarctica as common heritage—for exploitation	Fairly uncertain	3	Fairly important	3	**CU**	**9**
Level of support for Antarctica as common heritage—for preservation	Fairly certain	2	Fairly important	3	*PE*	*6*

In Column B, we score *importance* as follows:

- 1 = very unimportant
- 2 = fairly unimportant
- 3 = fairly important
- 4 = very important.

In Column C, we indicate whether the driving force is—predominantly—a *predetermined element* (**PE**) or a *critical uncertainty* (**CU**), based on the scoring from

Column A. Scores of 1 or 2 are classed as a predetermined element, and scores of 3 or 4 are classed as a critical uncertainty.

Finally, in Column D, we multiply the uncertainty score by the importance score to give an overall score, with possible total scores ranging from 1 to 16. The highest overall scores will give us our top critical uncertainties, and we will select two of these to construct our IL 2 × 2 matrix.

The driving forces for militarisation identified here are illustrative and not exhaustive. Other scenario analysts are likely to compile different lists and arrive at different ratings. A rigorous, 'real world' scenario process would allow sufficient time for different perspectives to be considered, and for stakeholder feedback to be sought from government, industry, civil society, and international organisations. We note that, for reasons of security or sensitivity, information sharing may be constrained by confidentiality requirements. For this reason, we suggest that a two-layered engagement process be undertaken. This would enable organisations to manage secure or sensitive information internally, while also gaining valuable insights from a broader group of stakeholders. Ideally, several iterations of the driving forces will be produced and reviewed by participants prior to developing the final IL 2 × 2 matrix and scenario narratives.

We now turn to the driving forces for Antarctic militarisation across the nine themes categories identified in Chap. 5: political, economic, social, technological, environmental, legal, geographic, strategic, and historical.

This assessment is for demonstration purposes only, and we will not justify each of the scores given here. However, in a 'real life' exercise, it is crucial that we have vigorous conversations to test our assumptions and assessments of each driving force. Investing time at this stage will help us to develop scenarios that are plausible, rigorous, and robust.

Table 7.2, identifies which of the driving forces of militarisation can be classed as critical uncertainties (CUs). Of our critical uncertainties, the following received maximum scores of 16, being assessed as both *very uncertain* and *very important*. We label these 'CU 1' and 'CU 2' (Table 7.3).

In addition, the following critical uncertainties received scores of 12, being assessed as both *fairly uncertain* and *very important* (Table 7.4).

Many of these concern technological developments in GPS-style systems, military satellite communications and infrared telescopes. We view many of these issues as closely related to the strategic advantages of Antarctic militarisation (CU 2). Similarly, we note that adherence to Treaty obligations is closely related to the stability of the international system (CU 1). By joining related critical uncertainties in this way, we can broaden the number of matters we consider within a single scenario, while preserving the basic logic of IL 2 × 2.

Table 7.3 Top two critical uncertainties for Antarctic militarisation

Political	Stability of the international system, relevant to the ATS	CU 1
Strategic	Strategic advantage from Antarctic militarisation	CU 2

Table 7.4 Other significant critical uncertainties for Antarctic militarisation

Technological	Use of ground receiving stations for military purposes
	Use of links to military satellites from ground stations
	Use of infrared telescopes for military purposes
	Use of unmanned aircraft systems for military surveillance and offensive capability
Environmental	Severity of climate change impacts
Legal	Adherence by Parties to Treaty obligations relating to Antarctica
	Use of civilians for military purposes
	Capability of inspection regime to detect dual use of equipment
Geographic	Extent of ice on land and sea

In summary, based on our subjective assessment, the critical uncertainties with the highest levels of importance and uncertainty are CU 1 and CU 2. These concern the future stability of international political systems relevant to the ATS, and the strategic utility of conducting non-peaceful military activity within Antarctica and the Southern Ocean. As outlined in Chap. 4, the *interaction* of these two critical uncertainties will define the parameters for the scenarios we construct. We will return to this in Step 6, below. Finally, even though we will not use any of the predetermined elements identified above to construct our IL 2 × 2 matrix, some of them may still exert a significant influence on how a scenario unfolds. For this reason, the predetermined elements should not be discarded, but instead retained for consideration when building our scenario narratives.

7.7 Step 6: Select Your General Approach/es

In Chap. 4, we saw that if we adopt a *deductive* approach to our scenarios, we *work backwards* from a future state defined from a set of premisses we identify at the beginning of the process. Then, drawing on human intuition, we construct a plausible narrative to describe a logical sequence of events that could take us from the present to that future state. We also noted that that our premisses may need to evolve as the scenario unfolds, and new information comes to light. Deductive premisses can take different forms, depending on the technique we use. They may be *driving forces* of the sort we have identified here, as in Step 1, or *archetypes* (idealised models), such as utopia and dystopia, or other generic images of the future. Deductive premisses may also be *normative* propositions, which describe a desired future state and assume that we have agency to make choices that take us toward that preferred state. Whatever the source of our premisses, with a deductive approach we work backwards from a defined future state towards the present.

In contrast, if we adopt an *inductive* approach, we *work forwards* in time, starting from analysis of current events or variables, proceeding step-by-step and without any preconceived idea of where things may head. Here, too, our inductive reasoning will be based on comprehensive data, rigorous analysis, expert judgement, imagination, and intuition to validate the plausibility of the sequences of events that form the bones of our scenario narrative.

So, which approach should we choose for our scenarios on Antarctic militarisation? As we noted in Chap. 4, both have their strengths and weaknesses. Induction allows us to work with a much larger number of driving forces in a scenario than, for example, the deductive IL 2 × 2 approach, which is focused primarily on the interplay of just two. We also noted that the open-ended nature of induction may allow more room for us to think creatively and mount plausible challenges to our current preconceptions and assumptions, thus generating novel and surprising insights. On the downside, an inductive process can be harder to facilitate and scope, and it can be harder to achieve internal coherence.

Ideally, we would recommend adopting *both* approaches, to draw on their strengths and to compensate for their flaws. This could be achieved simultaneously by two separate teams, working from the same base of information. Here, we will follow the (deductive) IL 2 × 2 approach, as detailed in Chap. 4. We choose this technique primarily because more comprehensive alternatives are too lengthy to cover here. However, we acknowledge the limitations of this approach.

7.8 Step 7: Construct the Scenario Matrix

In this step, we take our top two critical uncertainties (identified in Step 5) to form a 2 × 2 matrix. The matrix will define the logic and boundaries for four contrasting, plausible futures. Our two critical uncertainties are set out in Table 7.5.

As we have seen, our critical uncertainties are selected, in part, because they can occur (plausibly) at low and high values. In this case, it is plausible that, within our time horizon of ten years, the international system could be either quite stable or quite unstable. It is also plausible that there could be either considerable or negligible strategic advantage in states undertaking non-peaceful military activity in Antarctica.

Table 7.5 Top two critical uncertainties for the scenario matrix

Stability of the international system, relevant to the ATS	Stability of the international system may decrease, with a shift to multipolarity and great power competition, particularly between China, Russia, and the United States, with potential for spill over effects into the ATA	CU 1
Strategic advantage from Antarctic militarisation	Level of military / strategic advantage from use of Antarctica and the Southern Ocean for non-peaceful activities such as military research and substantive military activity	CU 2

Before we go on to construct the scenario matrix and draft the scenario narratives, it is worth pausing to consider the 'dynamic signatures' of our top two critical uncertainties. In Chap. 5, we observed that not all the driving forces of geopolitics operate in the same way within our scenarios: (1) some will tend towards being critical uncertainties, and others will be predetermined; (2) they may impel events or trends forward or constrain them; (3) they may be exogenous or endogenous to the field in question; and (4) they will tend to instigate rivalry or promote cooperation. We concluded that it is useful to represent each of these dynamics as a spectrum, rather than a binary choice between two extremes. And we proposed that the combination of these four spectra can be represented graphically as the 'dynamic signature' for a given driving force. At this point, we ask: is the signature of a critical uncertainty different when it occurs at its high value, compared to its low value? And finally, could these signatures change (in important ways) within the time horizon of our scenario?

We suggested that unpacking our driving forces in this way provides a more nuanced understanding of the causal impacts they are likely to have within a scenario. This can be useful in two ways. First, it may help clarify (and, if necessary, challenge) the assumptions that scenario participants may hold about the impact of a particular driving force. Second, it can help ensure that the decisions we make about the trajectory of a scenario narrative are internally consistent with our assumptions.

Here, we suggest a possible dynamic signature for each critical uncertainty, in their low and high instances.

In Fig. 7.1, we suggest that:

- The stability of the international system, relevant to the ATS (CU 1) is largely *exogenous* to the ATS.
- Where stability is high, the *constraints* on competitive behaviour within the ATS are likely to be effective and will promote *cooperation* among Parties.

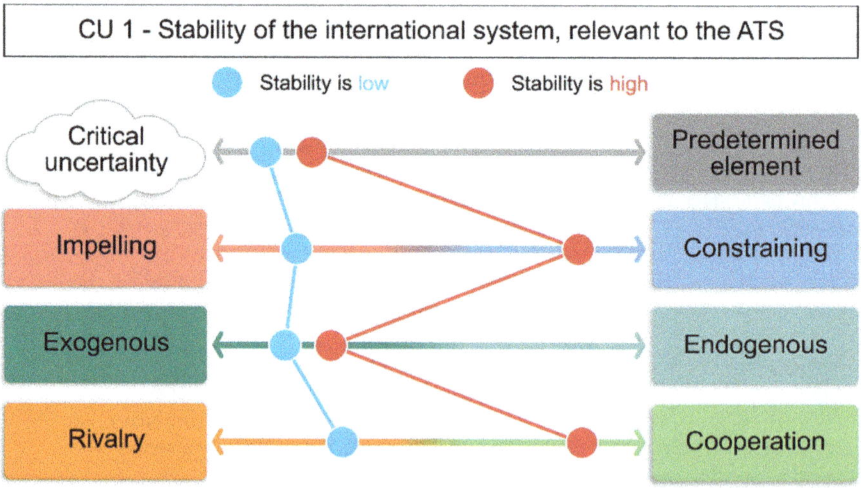

Fig. 7.1 Dynamic signature of CU 1 at and low and high instances

Table 7.6 An IL 2 × 2 scenario logic for Antarctic militarisation

Scenario title	Combinations of CUs at high and low values	Outcome
Splendid isolation	• CU 1—international stability is high • CU 2—advantage of militarisation is low	Cooperation Cooperation
Quarantine	• CU 1—international stability is low • CU 2—advantage of militarisation is low	Rivalry Cooperation
Stealth	• CU 1—international stability is high • CU 2—advantage of militarisation is high	Cooperation Rivalry
Cold War II	• CU 1—international stability is low • CU 2—advantage of militarisation is high	Rivalry Rivalry

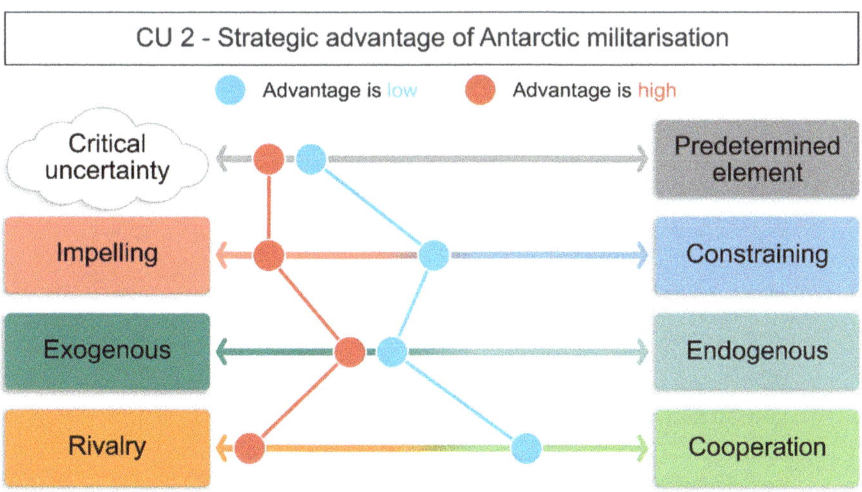

Fig. 7.2 Dynamic signature of CU 2 at low and high instances

In Fig. 7.2, we suggest that:

- The strategic advantages of Antarctic militarisation (CU 2) are primarily *exogenous* to Antarctica (focused on systems to support military capabilities elsewhere); however, there is also a smaller, *endogenous* aspect: it is at least plausible that military activity could be undertaken by Parties (in breach of the Antarctic Treaty) or by third parties, to assert or protect strategic interests within the ATA itself.
- The high and low instances of this critical uncertainty have a profound impact on the probability of rivalry or cooperation in the area below 60 degrees South.

Having identified our top two critical uncertainties and noted some of their dynamics, we are ready to assemble the scenario logic. Using the IL 2 × 2 technique, if we combine the two critical uncertainties (CU 1 and CU 2) and their two values (high and low), we can construct four distinct scenarios, which we name 'Splendid isolation', 'Quarantine', 'Stealth' and 'Cold War II' (Table 7.6).

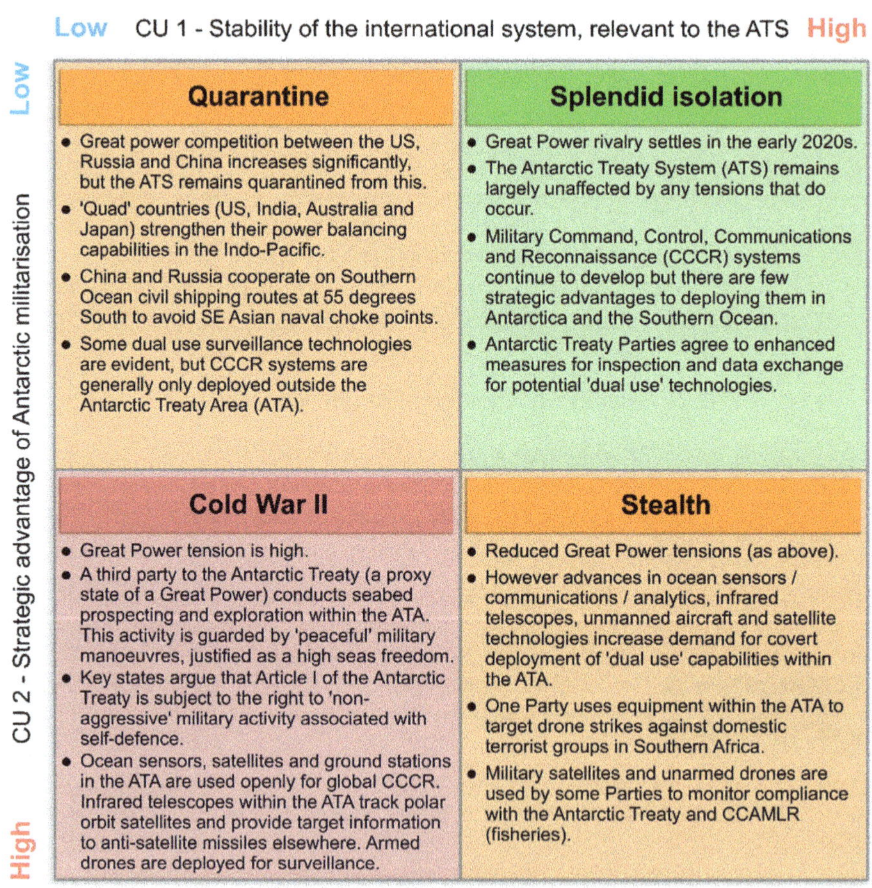

Fig. 7.3 Four scenarios of Antarctic militarisation

Each of these scenarios is placed within the matrix, with bullet points describing its key features (Fig. 7.3).

Having constructed the matrix, our next step is to craft more detailed narratives for each of these scenarios. For this, we turn to Chap. 8.

7.9 Summary

In this chapter, we have illustrated a process for constructing scenarios on Antarctic militarisation in the classical geopolitical tradition. We have drawn on our findings (in previous chapters) about Antarctic geopolitics in general, and about militarisation in particular. Importantly, we have augmented the standard IL 2 × 2 process in several ways to address some of the difficulties encountered when applying this technique to Antarctic geopolitics. We followed the nine basic steps of the IL 2 × 2 scenario technique, set out in Chap. 4. We acknowledged the limitations of this exercise that would be overcome in a 'real world' process, with a broader set of participants and multiple iterations.

The scenario standpoint we adopted (Step 1) is that of states whose policy and actions preserve and enhance the peaceful use of Antarctica and the Southern Ocean. We set the time horizon for these scenarios (Step 2) at around a decade. We identified around 50 driving forces relevant to Antarctic militarisation (Step 3), grouped across nine themes. We then classified each of these driving forces as (predominantly) either a *predetermined element* or a *critical uncertainty* (Step 4). Next, we ranked the critical uncertainties based on *importance* and *uncertainty* (Step 5). We adopted a *deductive* approach to these scenarios (Step 6), where we work backwards from a future state that is defined from a set of premisses we select at the beginning of the process. Here, the premisses are our two top-scoring critical uncertainties. These critical uncertainties form the two axes of our IL 2 × 2 scenario matrix (Step 7).

Political	Stability of the international system, relevant to the ATS	CU 1
Strategic	Strategic advantage from Antarctic militarisation	CU 2

We then offered an informed, but subjective, assessment of the *dynamic signature* of each of these two critical uncertainties (at the present time) at their high and low values. Finally, we sketched the outlines of four scenario narratives, based on the four quadrants of the 2 × 2 scenario matrix (Step 8). These quadrants reflect the four outcomes that arise when we combine our two critical uncertainties at their high and low values. These narratives suggest a range of plausible outcomes within our ten-year time horizon, from a highly cooperative scenario (*Splendid isolation*) to a highly rivalrous one (*Cold War II*).

We unpack these scenarios further in Chap. 8.

References

Allison, Graham. 2017. *Destined for War: Can America and China Escape Thucydides's Trap?* Boston, MA: Houghton Mifflin Harcourt.

Bateman, Sam. 2013. Is Antarctica Demilitarised? *The Strategist*, 23 April. Canberra: Australian Strategic Policy Institute. https://www.aspistrategist.org.au/is-antarctica-demilitarised/. Accessed 12 August 2020.

Bergin, Anthony, and Tony Press. 2020. Eyes Wide Open: Managing the Australia-China Antarctic Relationship. *Special Report* 153 (April 2020). Canberra: Australian Strategic Policy Institute. https://www.aspi.org.au/report/eyes-wide-open-managing-australia-china-antarctic-relationship. Accessed 5 May 2021.

Bergin, Anthony, David Brewster, François Gemenne, and Paul Barnes. 2019. *Environmental Security in the Eastern Indian Ocean, Antarctica and the Southern Ocean: A Risk Mapping Approach.* Canberra: Australian National University, National Security College/Paris: Institut de Relations Internationales et Stratégiques (IRIS). https://www.iris-france.org/wp-content/uploads/2019/06/Environmental-security-in-Eastern-Indian-Ocean-1.pdf. Accessed 28 July 2021.

Bieniek, Matthew, Dan Ellis, David Groce, Kerryn McCallum, and Alice Paton. 2020. *Post-COVID World: In the Post-COVID World, What Are the Possible Scenarios for a Post-COVID19 Global Order and What Effects May that Have on Australian Strategic Policy?* The Perry Group Paper, Team Sagan. Canberra: The Forge, Department of Defence. https://theforge.defence.gov.au/sites/default/files/2020-10/Post-COVID%20World_0.pdf. Accessed 28 July 2021.

Brady, Anne-Marie. 2017. China's Expanding Antarctic Interests Implications for Australia. *The Strategist*, 17 August. Canberra: Australian Strategic Policy Institute. https://www.aspi.org.au/report/chinas-expanding-interests-antarctica. Accessed 27 July 2021.

Brady, Anne-Marie. 2018. China in Xi's 'New Era': New Zealand and the CCP'S 'Magic Weapons.' *Journal of Democracy* 29 (2): 68–75.

Brands, Hal, and Charles Edel. 2019. *The Lessons of Tragedy: Statecraft and World Order*. New Haven: Yale University Press.

Buchanan, Elizabeth, and Ryan Burke. 2021. Introducing 'Project 6633'—A Program on Arctic and Antarctic Security. West Point, New York: The Modern War Institute (MWI) at West Point, Department of Military Instruction. https://mwi.usma.edu/introducing-project-6633-a-program-on-arctic-and-antarctic-security/. Accessed 29 July 2021.

Commonwealth of Australia. 2006. RAN Activities in the Southern Ocean. *Semaphore – Newsletter of the Sea Power Centre – Australia*, 18 October. Canberra: Sea Power Centre, Royal Australian Navy. https://www.navy.gov.au/sites/default/files/documents/Semaphore_2006_18.pdf. Accessed 28 July 2021.

Commonwealth of Australia. 2016a. Australian Antarctic Strategy and 20 Year Action Plan. Canberra: Commonwealth of Australia. https://www.antarctica.gov.au/about-us/antarctic-strategy-and-action-plan/. Accessed 28 July 2021.

Commonwealth of Australia. 2016b. *2016 Defence White Paper*. Canberra: Department of Defence. https://www.defence.gov.au/whitepaper/Docs/2016-Defence-White-Paper.pdf. Accessed 28 July 2021.

Commonwealth of Australia. 2018. Department of Defence—Submission to the Joint Standing Committee on the National Capital and External Territories Inquiry into the Adequacy of Australia's Infrastructure Assets and Capability in Antarctica. Submission 14. Canberra: Commonwealth of Australia. https://www.aph.gov.au/Parliamentary_Business/Committees/Joint/National_Capital_and_External_Territories/AntarcticTerritory/Submissions. Accessed 28 July 2021.

Commonwealth of Australia. 2020a. Operation Southern Discovery - Antarctic. Canberra: Commonwealth of Australia, Department of Defence. https://www1.defence.gov.au/operations/southern-discovery. Accessed 28 July 2021.

Commonwealth of Australia. 2020b. *2020 Defence Strategic Update and 2020 Force Structure Plan.* Canberra: Commonwealth of Australia, Department of Defence. https://www1.defence.gov.au/strategy-policy/strategic-update-2020. Accessed 28 July 2021.

Haass, Richard. 2020. *The World: A Brief Introduction.* New York: Penguin Books.

Hodgman, Paddy. 2016. Some Capability and Operating Implications Arising from Australia's Antarctic and Southern Ocean Interests. *Soundings*, Paper No. 10. Canberra: Sea Power Centre, Royal Australian Navy. https://www.navy.gov.au/sites/default/files/documents/Soundings_No_10.pdf. Accessed 28 July 2021.

Huleatt, Michael B. 2019. Australian Resource Reviews: Rare Earth Elements. Canberra: Geoscience Australia.

Layton, Peter, Travis Hallen, and Lauren Bishop. 2019. *Australia's Antarctic National Air Power Futures.* Canberra: Air Power Development Centre, Department of Defence. https://airpower.airforce.gov.au/publications/australias-antarctica-national-air-power-futures. Accessed 28 July 2021.

Luce, Edward. 2017. *The Retreat of Western Liberalism.* London: Little, Brown.

Mearsheimer, John J. 2018. *Great Delusion: Liberal Dreams and International Realities.* New Haven: Yale University Press.

Mearsheimer, John J. 2019. Bound to Fail: The Rise and Fall of the Liberal International Order. *International Security* 43 (4): 7–50.

Medcalf, Rory. 2020. *Contest for the Indo-Pacific: Why China Won't Map the Future.* Melbourne: La Trobe University Press.

Mullerson, Rein. 2017. *Dawn of a New Order: Geopolitics and the Clash of Ideologies.* London: Bloomsbury Publishing.

Parsons, Anthony, ed. 1987. *Antarctica: The Next Decade. Report of a Group Study Chaired by Sir Anthony Parsons.* Cambridge: Cambridge University Press.

Press, Tony. 2019. Australia Wants to Install Military Technology in Antarctica—Here's Why That's Allowed. *The Conversation*, 23 August 2019. Melbourne: The Conversation Media Group. https://theconversation.com/australia-wants-to-install-military-technology-in-antarctica-heres-why-thats-allowed-122122. Accessed 12 August 2020.

Schwartz, Peter. 1996. *The Art of the Long View: Planning for the Future in An Uncertain World.* Chichester, United Kingdom: John Wiley & Sons.

Taylor, Brendan. 2018. *The Four Flashpoints. Asia Goes to War.* Melbourne: La Trobe University Press.

Tetlock, Philip E., and Dan Gardner. 2015. *Superforecasting: The Art and Science of Prediction.* New York: Broadway Books.

Van der Heijden, Kees. 1996. *Scenarios: The Art of Strategic Conversation.* Chichester, UK: Wiley.

Young, Claire. 2018. What's China Up to in Antarctica? *The Strategist,* 20 September. Canberra: Australian Strategic Policy Institute. https://www.aspistrategist.org.au/whats-china-up-to-in-antarctica/. Accessed 12 August 2020.

Young, Claire. 2021. Eyes on the Prize; Australia, China and the Antarctic Treaty System. Policy Briefs, 16 February. Sydney: Lowy Institute.

Chapter 8
Antarctic Militarisation: Five Scenarios

Abstract In this chapter, we build on the work of previous chapters to develop five plausible scenarios on Antarctic militarisation, using the deductive approach and the IL 2 × 2 technique. Our time horizon is around a decade into the future. The 'scenario standpoint' we adopt is that of states whose policies and actions seek to preserve and enhance the current rules and norms of the ATS—including promotion of the peaceful use of Antarctica and the Southern Ocean. Our first four scenarios are titled *Splendid Isolation*, *Quarantine, Stealth*, and *Cold War II*. Our fifth scenario is *Pressures Managed*—a new medium option. This is designed to provide us with some evidence-based, targeted, broad-brush prediction, without suggesting this outcome is more probable. From these five narratives, we offer a set of policy-relevant recommendations. If the non-militarisation of Antarctica is to be preserved, two suggestions stand out. The first is that like-minded states should resist moves to reinterpret what constitutes 'military activity' in Antarctica to a lesser standard. The second is that the ATS should review its compliance and inspection regimes, to better account for contemporary military technologies.

Keywords Peaceful use of Antarctica · China · Russia · United States · Antarctic rules and norms · CCCR systems · Dual-use equipment and personnel · Ocean of Things · High seas freedom of navigation and Antarctica

8.1 Introduction

In Chap. 7, we sketched the outlines of four scenarios on Antarctic militarisation using a deductive approach and an intuitive logics (IL 2 × 2) technique. In this chapter, we illustrate how these scenarios might unfold over the next ten years. In Chap. 3, we suggested that many organisations expect that their scenarios will deliver some form of *evidence-based*, *broad-brush,* but *targeted* prediction. For this reason, we will add a fifth scenario. To us, an indicator of success is that the realised future falls somewhere within the total basket of scenarios produced, when taken as a whole.

Before we start, it is useful to recap the objectives for our scenario narratives. In Chap. 3, we noted that scenarios are not just static portraits of some future point in time. Instead, they describe a plausible *sequence of events* (a causal pathway)

J. McGee et al., *The Future of Antarctica*, Springer Polar Sciences,
https://doi.org/10.1007/978-981-16-7095-4_8

that leads us from the present to some future state. We saw that to be plausible, our scenarios cannot be mere flights of fancy. Instead, their starting point should be sufficiently anchored in present events and trends. Then, we build each narrative through time, with the scope of our stories bounded only by the parameters of our matrix. We use imagination to challenge our assumptions, and to cast off the blinkers of incrementalism. At the same time, we use structured thinking to achieve a disciplined ordering of data and perceptions, and to ensure that each scenario is internally consistent. Finally, good scenarios lead us to reperceive familiar phenomena in new ways (however unsettling); they provide inputs to strategic planning; and they help organisations to 'stress test' their current plans. With these objectives in mind, we proceed to the first of our scenarios.

8.1.1 Scenario 1: Splendid Isolation

This scenario describes a future in which Antarctic geopolitics progresses in 'Splendid Isolation'. There is a low level of tension in the international system from great power competition, and a low level of strategic/military advantage in the use of the ATA.

The early 2020s saw significant and rising international tension between the great powers China, Russia, and the United States, during the profound loss of life and economic disruption caused by the COVID-19 pandemic. The more traditional US presidency of Joe Biden from early 2021, and successful COVID vaccination programs in most countries over 2022–24, settled the more severe aspects of the tensions between the great powers by late 2025. The Antarctic Treaty Consultative Meeting (ATCM) has been largely unaffected by this period of great power tension and rivalry in the first half of the 2020s. Consensus decision-making within the ATCM and CCAMLR meetings continues to function reasonably well, except that some proposals for marine protected areas and special area management on the Antarctic continent continue to be sticking points. However, the Parties to the Antarctic Treaty and CAMLR Convention have been able to isolate lack of consensus to those issues, so that the otherwise largely cooperative working relationship of states within the ATS has prevailed.

By the mid-2020s, technological developments in ocean sensors, satellite technologies, infrared telescopes, satellite ground stations and global positioning systems have advanced incrementally, meaning that their military applications for situational awareness and CCCR functions are moderately more attractive. However, these technological developments have not led to any significant increase in the advantages of developing such capacity in the ATA. States such as China and Russia, which have alternate networks to the US-controlled GPS system, have therefore invested in placing equipment in friendly developing countries located in the lower latitudes of the southern hemisphere (i.e., 35°–45° South), well outside the ATA. Equipment located in these temperate latitudes now provides similar performance to the equipment placed within the Antarctic area.

As allowed under the Antarctic Treaty, states have continued to use military personnel for logistical support of research bases and scientific work on the Antarctic continent and in the Southern Ocean. Activity by military personnel in this capacity continues to receive uneven reporting by the ATCPs; however, there is no strong evidence that any military personnel below 60° South have been engaging in military research or activities. In 2026, consensus was reached within the ATCM on a new transparency measure that will allow Antarctic Treaty states to request and receive 'macro data' evidencing the type and nature of all information received, transmitted, and produced from radio-telescopes, infrared telescopes, satellite ground stations, and ocean sensors located below 60° South. This new binding measure was agreed between the ATCPs due to concerns about the possible 'dual use' of space research facilities and ocean sensors; that is, short-term repurposing between civilian/scientific use and use in military CCCR systems.

In reaching consensus on this transparency measure, the ATCPs reaffirmed the importance of Article I of the Antarctic Treaty and its prohibition on activities of a military nature. The Parties were able to agree that the nature of military activities in the 2020s had broadened in scope from the use of military ships, troops, and aircraft in Antarctica anticipated by the drafters of the Treaty in the late 1950s. The potential for equipment such as infrared telescopes, radio-telescopes, ocean sensors, and ground station receivers to be used in a global network of military CCCR capacities was acknowledged by the Parties. This was the driving force behind efforts to have greater transparency in the use to which Antarctic research equipment is put. Leading states such as China, Germany, Russia, and the United States have requested macro data from infrared telescopes, ground station receivers and ocean sensors used in the ATA by other states. Analysis of this macro data has not yet raised any specific concerns regarding repurposing of such equipment within the ATA for use in military CCCR systems.

8.1.2 Scenario 2: Quarantine

This scenario describes a future in which Antarctic geopolitics is 'quarantined' from global tensions. There is high great power competition in the international system, but low strategic value in conducting military activities in the area below 60° South.

In the mid-2020s, great power competition between the United States, Russia, China, and India increased significantly due to tensions over maritime claims in the South China Sea, East China Sea, Northern Europe, and the Black Sea area. China and Russia have formed a more overt strategic relationship, and openly seek to change parts of the US-designed, post-World War II 'rules-based order', where such rules and norms conflict with their plans to keep open possibilities for resource extraction in areas such as the Arctic and Antarctica. The Indo-Pacific 'Quad' coalition of the US, India, Japan, and Australia has been significantly strengthened in the period to 2025 through regular US-led meetings of leaders and military cooperation/exercises in the Indian and Pacific Oceans. This has led to tense maritime stand-offs between

Russian and Chinese Coast Guard/naval vessels and Quad naval assets deployed in the South China Sea, Taiwan Strait and East China Sea, and the Black Sea area. Diplomatic efforts at resolving these tensions through international agreement have so far proved unfruitful, and efforts to initiate formal or informal arbitration or judicial settlement have been rebuffed. However, these naval tensions over maritime zones have not extended to the Southern Ocean or into the ATA.

While technologies relating to infrared telescopes, global positioning systems, satellite communications, ocean sensors, and unmanned aircraft and ships continue to develop and offer potential military application, they have no obvious advantages in terms of being located within the ATA—compared with locating them at more accessible latitudes of 35°–45° South. Importantly, there are no obvious or significant advantages in locating or using such equipment on the Antarctic continent.

However, from 2025 China and Russia have cooperated on planning and testing Southern Ocean civilian shipping routes at approximately 55° South, which can provide a civilian naval supply route between the Indian and Pacific Oceans that avoids the 'choke points' of the Malacca Straits (which could be vulnerable in times of heightened international conflict) and the Suez Canal (which is vulnerable to blockage—as occurred in 2021—or to terrorist activity). China, Russia, and the US have deployed several ocean-based sensors in the area between 45° and 59° South in the vicinity of these civilian shipping routes. There is concern among the near Antarctic states (Australia, New Zealand, Argentina, Chile, and South Africa) that these ocean-based sensors might have capacity beyond gathering scientific data on ocean conditions. There is also concern these ocean sensors might support military satellite communications and global positioning systems, as well as having capacity to undertake surveillance of the location of naval shipping and submarines in the Southern Ocean. Australia and New Zealand have been particularly concerned about these developments in the maritime approaches below their respective southern borders. They are considering deploying similar ocean sensors and using manned and unmanned naval vessels to routinely patrol areas north of the boundary of the ATA.

Aerial surveillance by unmanned, unarmed aircraft (with ranges up to 15,000 km per mission) and military satellites has become commonplace as an inspection process under the Antarctic Treaty. This has precipitated more extensive surveillance of fishing and other activities in the Southern Ocean. The US has led the way in this regard, using airports in New Zealand to launch inspection flights in East Antarctica. China has also developed the capacity to deploy unmanned, unarmed aircraft and military satellites through use of airports in friendly states in southern Africa and South America. There have been moves within CCAMLR to expand aerial and satellite monitoring. Supportive countries point to a similar right of aerial observation in Art VII(4) of the Antarctic Treaty. However, consensus has been unable to be reached to formally add such a new inspection modality for CCAMLR. The parties are continuing discussions and are particularly interested in how such technologies might address a new spike in IUU fishing activities (by non-CCAMLR states) in areas around sub-Antarctic islands in the Indian Ocean.

8.1.3 Scenario 3: Stealth

This scenario illustrates a future in which Antarctica is militarised by stealth. There is low international tension, but high strategic advantage in using equipment and personnel for military purposes within the ATA.

The first few years of the 2020s witnessed an increase in international tension between China, Russia, and the US amid the political fallout from the COVID-19 pandemic and the legacy of the Trump presidency. By 2023, however, the Biden presidency in the US settles relations between the great powers. New agreements between the great powers on COVID vaccine distribution to the developing world and emissions reduction on climate change opened avenues for new and more extensive cooperation on key international issues; these have helped ease tensions between the great powers, which flared during the early 2020s.

Antarctic Treaty system meetings have been largely unaffected by the period of great power rivalry in the early 2020s. Consensus decision-making within the ATCM and CCAMLR continues to function reasonably well, except for proposals for new marine protected areas, an expansion of the commercial krill harvesting into East Antarctica, and special area management on the Antarctic continent. The Parties to the Antarctic Treaty and CCAMLR have been able to isolate lack of consensus to those issues, so the otherwise cooperative workings of the ATS have largely prevailed.

By 2025, there have been significant technological developments in ocean sensors, ocean communications platforms, global positioning systems, and satellite technologies. The US 'Ocean of Things' (i.e., 'persistent maritime situational awareness over large ocean areas by deploying thousands of small, low-cost floats that form a distributed sensor network' (Commonwealth of Australia 2016) provides the capacity for monitoring and surveillance of civilian ships, naval ships, and underwater vehicles across vast ocean areas, including the Southern Ocean. These developments enable improved situational awareness for military activities and CCCR systems.

As areas of the Indian, Pacific, and Atlantic Oceans above 60° South will be brought within the scope of the US 'Ocean of Things', there is increasing political and technological pressure to include the area below that latitude to make the system global. Other great powers, such as China and Russia, are developing their own version of the 'Ocean of Things'. China and the US have adopted a similar (but independent) position that the Antarctic Treaty restriction on military activities in ocean areas within the ATA does not affect the high seas freedom of navigation that allows 'peaceful' use of the region by naval forces. Both China and the US claim this interpretation of Articles I and VII of the Antarctic Treaty is consistent with the United Nations Law of the Sea Convention 1982 (UNCLOS), which allows for 'non-aggressive' naval navigation in all the world's oceans. Other Antarctic Treaty states, led by Australia, have argued strongly that Article I of the Antarctic Treaty provides for non-militarisation of the ATA, so that non-aggressive naval use of the oceans surrounding the Antarctic continent is prohibited, unless it is in support of Antarctic science and/or logistics. China and the US do not accept this interpretation, claiming that the high seas freedom of naval navigation under UNCLOS post-dates the Antarctic Treaty, and that Article VI of the Treaty provided for such developments in maritime law.

China and the US therefore maintain that there are effectively two zones of restriction of military activity within the ATA. In the oceans surrounding Antarctica, high seas freedoms allow for freedom of navigation of naval ships and freedom of overflight of military aircraft, if such military activities are of a 'non-aggressive' nature, as required more broadly by the UN Charter (1945). However, on the Antarctic continent itself, high seas freedoms do not apply, so no military activities are allowed, whether 'aggressive' or not. China and the US therefore intend to proceed with the placement of ocean sensors and ocean communication platforms in the Southern Ocean, in the area between the Antarctic coastline and 60° South which, in their view, are 'high seas areas'. This is despite the protestations of the Antarctic claimant states, which have asserted territorial seas and exclusive economic zones off the Antarctic coastline. States with sub-Antarctic islands have also voiced concern that such sensors and communication platforms, if tethered to the ocean floor, may infringe their rights of control over extended continental shelves.

In 2027, evidence comes to light of communications facilities on the Antarctic continent being used to relay military communications to satellites, which inform unmanned aircraft being used in southern Africa for armed strikes against domestic terrorist groups. The Antarctic communications facilities used for this purpose are operated by an Antarctic Treaty country, which claims this activity is with the consent of the southern African state using the unmanned aircraft (which is not an Antarctic Treaty state) and that it is within the bounds of Article 1 of the Antarctic Treaty, as it is not engaging in any internationally aggressive behaviour which would be in breach of the UN Charter. The Antarctic Treaty Party which operated the communication facilities claims that the restriction on 'military activities' in Article 1 of the Antarctic Treaty must also be interpreted in the context of similar references to 'peaceful use' of the oceans found in the 1982 Law of the Sea Convention and 'peaceful use' of space in 1967 Outer Space Treaty. These treaties having been long understood as allowing for non-aggressive activity in high seas areas and space, thereby allowing for these areas to be used for military communications and surveillance.

8.1.4 Scenario 4: Cold War II

This scenario describes a return of militarisation to Antarctica. It illustrates a plausible (and alarming) future in which there is high international tension between the great powers, and high strategic advantage in states conducting military activities in Antarctica and the Southern Ocean.

In the early 2020s, great power competition between China, Japan, India, Russia, and the US increases significantly due to tensions over maritime claims in the South China Sea and East China Sea, Northern Europe, and the Black Sea. Russia and China form a more overt strategic coalition seeking to overturn parts of the US-sponsored 'rules-based order' where they conflict with their interests for greater resource extraction in the Arctic, Antarctica, and deep seabed. The 'Quad' partnership of Australia, India, Japan, and the US expands to include Canada, Indonesia, New Zealand, and Singapore.

By 2025, this 'expanded Quad' has been strengthened through regular diplomatic meetings and military cooperation exercises with a view to providing a wide-ranging coalition of states in the Indo-Pacific region that might balance the power of China, which has rapidly expanded its navy to 450 ships and achieved significant numerical and operational advantage over US naval assets in Southeast Asia and the Indian Ocean. European states, such as Germany and France, have also participated in naval exercises and freedom of navigation voyages with the Quad states.

There are tense stand-offs between Chinese Coast Guard/naval vessels and Quad naval assets in the South China Sea, Taiwan Strait, and East China Sea, and between Russian and the North Atlantic Treaty Organization (NATO) naval forces in the Black Sea and North Sea areas. Diplomatic efforts at resolving these tensions have so far proved fruitless, and efforts at any form of arbitration or judicial settlement have also been unsuccessful. Freedom of navigation patrols by Quad naval assets in the South China Sea and East China Sea have led to several near-miss incidents with Chinese Coast Guard and naval vessels.

In 2027, competition between the great powers over sourcing strategic minerals becomes intense due to their increased use in advanced weapons, communications equipment, and low carbon energy systems. China and Russia therefore encourage a southern African state, which is not a party to the Antarctic Treaty or CCAMLR, to put forward plans for seabed exploration and mining of cobalt in the high seas areas below 60° South. This southern African state argues that the deep seabed mining provisions in Part XI of the Law of the Sea Convention will not apply to this activity. This because the parties to the 1982 Law of the Sea Convention intended to exclude the region below 60° South from operation of the Convention, as the Antarctic Treaty Parties were negotiating a separate minerals arrangement at the time.

The US, Australia and many other 'likeminded' countries claim that, although this southern African country is not party to the Madrid Protocol, the mining prohibition is now customary international law, so is binding on all states. The southern African state has refused to enter any arbitration or conciliation of the issue and is pressing ahead with a close logistics and military relationship with China and Russia, which plan to provide 'technology transfer' in the form of use of civilian ships to assist with transport of personnel and equipment for the seabed mining. Key states in the international system, including China and the US, also openly contend that Article 1 of the Antarctic Treaty does not restrict high seas freedoms, which include non-aggressive naval activities in the area below 60° South. China and Russia are also relying upon this argument to justify their civilian logistical support for the mining operations of the southern African state.

During a routine inspection of equipment at a prominent Antarctic research base, the inspection process detects that radio telescopes and infrared telescopes are being used to track the location of polar-orbiting satellites of other states. This has been set up with a view to providing targeting information to military CCCR capacities of the military forces of this state, including information that might be used to shoot down or disable the polar satellites of other states. The military forces of this state are located outside the ATA, but communications from the radio telescopes and infrared telescopes at this base can be relayed to terrestrial missile batteries in the northern hemisphere, and/or killer satellites in orbit, which are designed to shoot down, or disable, the satellites of other states.

By 2025, a view has developed amongst some great powers that considering contemporary international practice regarding the 1982 Law of the Sea Convention and the 1967 Outer Space Treaty, 'peaceful use' of Antarctica needs to be understood with regard to practices in other international treaties. These powers argue that non-aggressive military activity is allowed even on the Antarctic continent, so long as force is not exercised in breach of the UN Charter. These great powers claim that the restrictions on military activity in the Antarctic Treaty have always been subject to a state's inherent right of self-defence under Article 51 of the UN Charter. This would allow for military (or civilian) equipment and personnel located in the ATA to be used in defence of the nation, when the nation was under imminent or actual attack, even if such attack occurs outside the ATA.

This view has triggered an upswing in the scale and type of 'dual-use' equipment being used at Antarctic research bases. This has led to a more open position that infrared telescopes, ground station receivers, ocean sensors and other similar facilities located on the Antarctic continent and in the Southern Ocean might be used for communications of a defensive military kind. These include CCCR facilities intended to defend the interests of the state located outside the ATA.

Finally, consistent with this interpretation of Article I of the Antarctic Treaty, some states have used armed unmanned aircraft and armed unmanned underwater vehicles for inspection and surveillance activities in the Southern Ocean and on the Antarctic continent. There has been no use of weapons so far, but they are being carried in an open manner on these unmanned aircraft and underwater vehicles.

In 2028, a civil society organisation from a prominent Antarctic Treaty country decided to carry out a visual inspection of the undersea activities of the southern African state by use of an unmanned, unarmed underwater vehicle, to gather information on the level of environmental harm that may be occurring to the seabed and benthic ecosystems. The southern African country responded by deploying an unmanned, armed underwater vehicle (thought to have been supplied by either China or Russia) which has disabled the first-mentioned underwater vehicle by firing and hitting it with a small torpedo. The US, Australia and other 'likeminded' countries have strongly criticised this action of the southern African state as an aggressive act in breach of the UN Charter. They also claim customary international law has incorporated the norm of non-militarisation of the area below 60° South and so is legally binding on all states, regardless of whether they are party to the Antarctic Treaty. These states have also criticised China and Russia for supplying the armed underwater vehicle used by the southern African country in the incident, although they stopped short of asserting that China and Russia have breached Article I of the Antarctic Treaty. In a heated diplomatic exchange, the southern African country, China, and Russia have rejected these arguments.

8.1.5 IL 5: A Fifth Scenario for Antarctic Militarisation

In Chap. 4, we argued that, while the IL 2×2 technique produces four well-contrasted outcomes, *if* our scenarios are to provide targeted, broad-brush prediction, then a middle option may be needed. We suggested the addition of a fifth scenario (in the centre), in a model we call 'IL 5'. Importantly, this fifth scenario is *not* meant to indicate the most probable outcome. Why? Because, as we saw in Chap. 3, scenarios become useful in the zone beyond standard prediction, and, in this zone, probability is hard to estimate. So, instead, our fifth scenario illustrates a medium 'expectation value', being the rough average of all possible outcomes, weighted by their likelihood. Here, our two critical uncertainties meet at their intermediate values, rather than at their extremes. This is illustrated in Fig. 8.1.

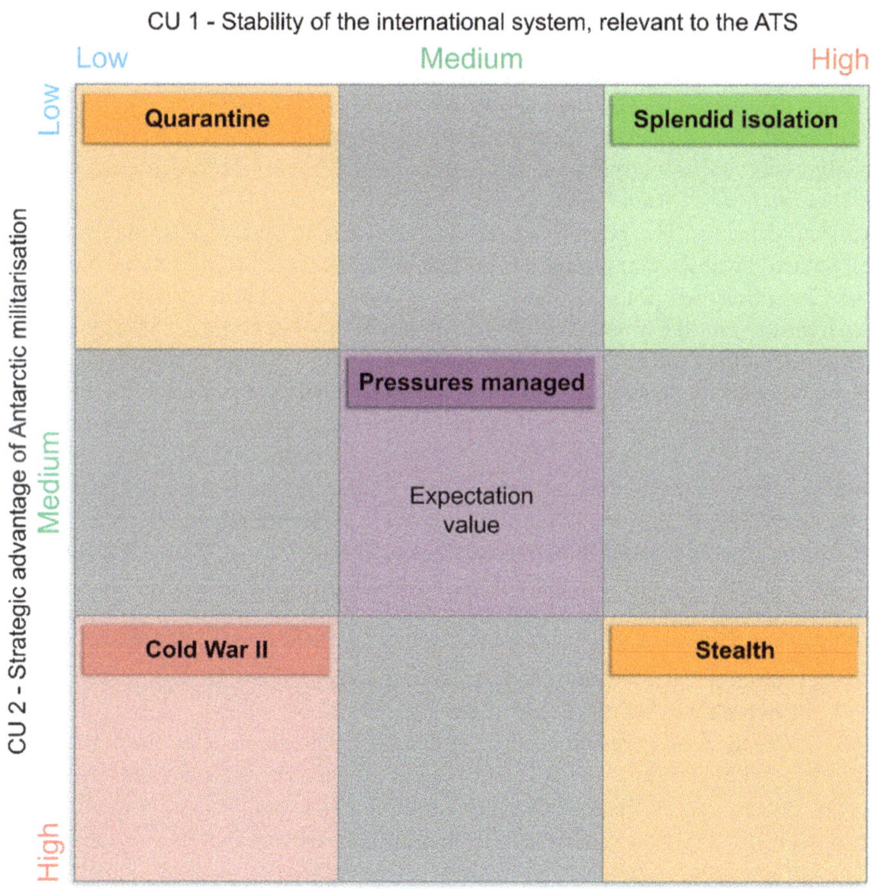

Fig. 8.1 An 'IL 5' model for Antarctic militarisation

8.1.6 Scenario 5: Pressures Managed

This final scenario provides some evidence-based, broad-brush, but *targeted* predictions *for Antarctic militarisation. It is based on the combination of our two critical uncertainties at their medium values. In this scenario, the various driving forces for Antarctic militarisation are managed more-or-less successfully. Scenario 5 is intended to be an anchor-point for pragmatic discussions on this issue. It should not eclipse the other scenarios, which remain equally plausible.*

At a global level, the early to mid-2020s witnessed a significant return to great power competition in the international system between China, Russia, and the US. This competition was heightened during the COVID-19 outbreak, the difficulties in the global economy which followed, and by the lack of trust in US foreign policy caused by the legacy of the Trump Administration. From 2021, the Biden Administration adopted a more measured tone to the issue of great power competition in the international system, but continued to sharpen American focus on a changed global political situation, in which the US is no longer the sole great power with the ability to shape the international order and the international law system in accordance with its interests. Over the period 2022–2025, this great power competition has been most keenly observed in international trade, through a US attempt to decouple various strategic industries from China in order to rebuild domestic industrial and manufacturing capacity. The great power competition has also been most manifested in heightened naval tension around areas such as the Taiwan Strait, South China Sea, East China Sea, and potential naval choke points in the Malacca Straits and Black Sea areas of central Europe. The US has joined with other countries in a Quadrilateral partnership in the Indo-Pacific region that is designed to act as a strategic power balancing coalition to check the rise of Chinese and Russian power in the region.

However, as with previous periods of great power competition between the US and Soviet Union during the 1950s and 1960s, cooperation in Antarctic affairs in the area below 60° South has not been significantly impeded by these wider tensions in the international system. The ATS has provided a forum where the great powers can engage in cooperation without significant loss of face or strategic advantage. By 2024, China has constructed a fleet of specialised krill-harvesting vessels and has commenced annual krill harvesting off the Antarctic peninsula area and East Antarctica; however, this is occurring at levels well below the catch limits for the relevant sub-areas. Russia has also re-entered the krill fishery, but again at a modest level of catch allowed by the CCAMLR.

By 2026, there has been an upsurge in demand for strategic Rare Earth Elements due to a significant shift towards greater use of electric vehicles and renewable energy systems, and for high-tech supercomputing and weapons systems. However, the geographic remoteness and cost of extraction of seabed minerals (such as cobalt) from seafloor areas of the Southern Ocean has remained economically unattractive, as more accessible, and less costly, deposits have opened in terrestrial reserves at high latitudes outside the ATA. However, Russia is engaged in voyages with survey vessels each year—which they categorise as ‘research activity’—that are directed at gathering information on the possible presence of natural gas and oil reserves and strategic minerals in offshore areas of East Antarctica and the Kerguelen Plateau. While government sources within Russia are silent on the issue, there is growing chatter amongst scientific and mining interests in Russia about the possibilities of mining in Antarctica after the Madrid Protocol comes up for a possible Review Conference from 2048 onwards.

The annual ATCM and meeting of CCAMLR have continued to respond to governance issues within the Antarctic region, after the less-productive online meetings during the COVID-interrupted period of 2020–22. Climate change has acted as a focal point for new research initiatives on Antarctic science between major global powers, allowing for an important part of international life to be less conflictual, and demonstrating the possibility of cooperation between the great powers, even in times of wider international tension.

During the early 2020s, there was a significant upturn in national investment in Antarctic science and logistics infrastructure. Several key Antarctic states began upgrades of scientific research bases and construction of transport infrastructure, such as larger research stations, new icebreakers, and improved air access. The planning, environmental assessment, and construction of this improved Antarctic science and logistics infrastructure has necessitated greater participation of military personnel at national Antarctic program bases. The states involved in this infrastructure have declared the role of these military personnel in reporting on their upcoming activities under the Antarctic Treaty. These military personnel have been engaged in surveying, geotechnical investigations and construction activities relating to this new infrastructure. There is no evidence to suggest increased military personnel are involved in any activities beyond this support of improved Antarctic infrastructure. However, by 2026, the frequency of inspections of Antarctic bases had risen by approximately 25% as ATCPs acknowledge the increased presence of military personnel on the continent and the importance of compliance with Article I of the Antarctic Treaty (non-militarisation of the Antarctic continent).

By 2027, technological developments in ocean sensors had provided new scientific, communications and strategic incentives for states to deploy these devices in areas of the Southern Ocean below 60° South. Ocean sensors gather scientific data on ocean conditions, but also act as a communications link for satellite communications and global positioning systems. Ocean sensors are therefore viewed by states as a possible part of a global military CCCR network, which each of the great powers is developing. China and the US have deployed such sensors in areas of the Southern Ocean above and below 60° South. Although both countries claim these sensors are primarily for the peaceful use of gathering scientific data, there is a clear potential dual-use capability would allow them to be repurposed at short notice for use in military CCCR networks.

In 2028, evidence emerges that infrared telescopes on the Chinese and US research stations located on the Antarctic continent have been used in tracking military satellites with polar orbits as they cross the Southern Ocean. China and the US claim the satellites are located above 100 km from the Earth's surface, so are outside the airspace of the ATA. They also claim such tracking activity is a 'non-aggressive' use of the ATA, so not in breach of Article 1 of the Antarctic Treaty. The issue has been introduced to the ATCM for clarification of whether such use of infrared telescopes is a breach of the Antarctic Treaty; however, as operators of the satellites (and Treaty Parties), China and the US can block consensus on this issue.

8.2 Step 9: Wind Tunnel Test Current Strategy and Prepare Policy Advice

Our final step involves the analyst formulating strategic advice to stakeholders based on the scenarios that are being constructed. At Step 1, the scenario standpoint we adopted was that of states whose policy and actions preserve and enhance the current rules and norms of the ATS. This includes promoting the peaceful use of Antarctica and the Southern Ocean. It is beyond the scope of this exercise to assess the robustness of current state strategies in light of these scenarios—a process often referred to as wind tunnel testing. It is also not possible to provide a full set of strategic recommendations, which would need to be tailored to the specific circumstances of individual states. Nonetheless, some key takeaways from our fivescenarios could be useful in formulating this advice:

1. **Key drivers for militarisation in Antarctica are *exogenous*to the ATA**. The key drivers for Antarctic militarisation in the decade ahead (i.e., global great power competition and technological developments increasing the value of military applications below 60° South) originate from outside the ATA and from outside the core functions and decision-making of the ATS.
2. **Insulate the ATS from international tension**. The ATS has traditionally been able to ameliorate the impact of broader global geopolitical tensions between great powers and act as a forum for international cooperation during difficult times. There is no guarantee that this relative isolation of the ATS will continue, however. As the world enters a new period of great power competition, it is important to bear this history in mind and make efforts to resist international tensions at higher latitudes spilling over into the ATCMs and meetings of CCAMLR.
3. **Focus on scientific cooperation**. A focus on Antarctic science and the global benefits of climate change science may provide a useful anchor-point for great power cooperation in Antarctica, which can serve as an example of the ability of great powers to cooperate on issues of mutual interest that are not deeply connected with the strategic issues dominant at higher latitudes.
4. **The nature of 'military activity' has changed**. The Antarctic Treaty was formed in the 1950s, when militarisation was likely to involve the placement of military personnel, bases, submarines, or intercontinental missiles in the area below 60° South. The inspection system under the Antarctic Treaty was therefore established to detect this type of militarisation of the Antarctic continent.

Over the last 60 years, however, development in technologies relating to satellites, telescopes, and ocean sensors has created a growing risk that great powers will look to develop globally networked military CCCR systems that involve use of assets located within the ATA.

5. **Antarctica is becoming more strategically attractive for the use of military technology**. Technological developments have the potential to make the placement and operation of satellites, telescopes, and ocean sensors in the ATA more attractive than in past decades. This is particularly the case with facilities that might be seamlessly repurposed between scientific activities within the ATS, and transmission of information with military applications relating to issues outside the ATS.
6. **Resist moves to reinterpret what constitutes 'military activity' to a lesser standard**. These technological developments lie at the intersection between the non-militarisation provisions of the Antarctic Treaty and similar provisions in the 1982 Law of the Sea Convention and the 1967 Outer Space Treaty relating to peaceful use of the oceans and space domain. The non-militarisation provisions of the Antarctic Treaty have traditionally been interpreted as providing for the highest level of restriction on military activities. We suggest that it is important to resist more liberal approaches to restrictions on military activity from influencing interpretive approaches to non-militarisation within the ATA.
7. **Review effectiveness of compliance and inspection regimes in light of new technologies**. Technological developments in satellites, telescopes, and ocean sensors, and the ability to seamlessly repurpose between civilian and military applications, represent a significant challenge to the effectiveness of the inspection regime within the ATS. This may require the development of new rules and processes, or at least some refinement of the interpretation of current rules, within the ATCM. However, these technological developments are probably best viewed as a double-edged sword: they might also create opportunities for greater use of satellites, telescopes, and ocean sensors in the inspection practices of the Antarctic Treaty and Madrid Protocol. Material gathered by such technologies might be useful for encouraging state compliance with the rules of the ATS, either as a source of public embarrassment for infringing states, or as evidence in any formal procedures that might be initiated for non-compliance.
8. **Consider invoking ATS dispute resolution mechanisms as a fallback option**. Aside from the possibility of initiating debate on the need for clarity in relation to Article I of the Antarctic Treaty in the context of these new technological developments, concerned states might also consider the possibility of an unprecedented threat to invoke the dispute resolution procedures of the Antarctic Treaty. Although invoking dispute resolution would seem an unlikely path in the current political context, it will remain a fallback option to challenge the legality of cases that push the boundaries of peaceful use of the ATA.

8.3 Summary

In this chapter, we unpacked five scenarios on Antarctic militarisation using the deductive approach and the IL 2 × 2 technique—see Chap. 4. Our time horizon is around a decade into the future. The basic premisses of our first four scenarios were established by combining our two top-scoring critical uncertainties: 'Stability of the international system, relevant to the ATS' (CU 1), and 'Strategic advantage from Antarctic militarisation' (CU 2)—see Chap. 7

In our first scenario, *Splendid Isolation*, a low level of tension in the international system combines with a low level of strategic/military advantage in the use of the ATA. To use the language of scenario archetypes (see Chap. 4), this is our *utopia.* Under these conditions, ATCPs agree to a new transparency measure that acknowledges the changing nature of military technologies and reaffirms the importance of the Antarctic Treaty's prohibition on military activity.

In our second scenario, *Quarantine,* there is high great power competition and tension, but low strategic value in conducting military activities in the area below 60° South. Under these conditions, a small number of powerful states challenge the rules and norms of the post-war international order—including the ATS. However, global tensions have not extended into the ATA. There are no significant advantages to using CCCR equipment on the Antarctic continent, though there is concern about dual use of marine-based sensors and unmanned, unarmed aircraft.

In our third scenario, *Stealth*, there is low international tension, but high strategic advantage in using equipment and personnel for military purposes within the ATA. Antarctic Treaty processes have been largely unaffected by great power rivalry, and the ATS continues to function—notwithstanding a lack of consensus over various measures for environmental protection. However, there is intense pressure to expand maritime surveillance systems, such as the 'Ocean of Things', into the ATA. At the same time, several major powers argue that Article I of the Antarctic Treaty does not preclude 'non-aggressive use' of Antarctic high seas by naval ships and military aircraft—and in their view, Antarctic 'high seas' encompass all seas below 60° South. Similarly, the 'non-military' concept is being re-interpreted to mean 'non-aggressive', to justify use of military communications and surveillance on the continent itself.

In our fourth scenario, *Cold War II*, there is high international tension between the great powers, and high strategic advantage for states to conduct military activities in Antarctica and the Southern Ocean. This is our *dystopia.* 'Dual-use' equipment is deployed increasingly at Antarctic stations. Several major powers use a proxy state to pursue deep seabed mining within the ATA, in possible contravention of the mining ban. This activity is guarded by military equipment and personnel, culminating in an armed attack against an unarmed civilian vessel. The long-standing non-militarisation of Antarctica is unravelling fast.

We have augmented these first four scenarios with a fifth scenario. This is because we believe that many organisations expect that scenarios will deliver some form of *evidence-based, broad-brush, but targeted* prediction in the zone beyond the limits of standard prediction—see Chap. 3. This fifth scenario is constructed on the premiss that our two critical uncertainties (CU 1 and CU 2) may occur, quite plausibly, at their medium values, rather than at their extreme high or low values. In Chap. 4, we noted that this fifth scenario is *probabilistically in the middle*, rather than something that is *more probable*.

In our fifth scenario, *Pressures managed*, there are considerable tensions in the wider international system, but these are kept in check by a new regional balance of power. Importantly, these tensions do not significantly impede state cooperation in Antarctic affairs. Major powers observe the catch limits for krill fisheries set by CCAMLR. Mining within the ATA remains uneconomic, but at least one power is exploring the potential for mining in Antarctica after the Madrid Protocol is open to a possible Review Conference from 2048. The shared challenges of climate change become a unifying point among Antarctic states and instigate a renewed scientific effort within the ATA. However, this is offset by steady growth in dual use technologies; the reinterpretation of 'non-military' activity as meaning 'non-aggressive'; and use of the ATS consensus rules to block tighter controls on dual-use technologies. Under this scenario, states wishing to preserve Antarctica for peace and science should be highly alert—but not yet alarmed.

From these five scenarios, we developed some policy-relevant recommendations, while acknowledging that comprehensive advice would be tailored to the circumstances of individual states. If the non-militarisation of Antarctica is to be preserved, two of these suggestions stand out. The first is that like-minded states should resist moves to reinterpret what constitutes 'military activity' in Antarctica to a lesser standard. The second recommendation is that the ATS should review its compliance and inspection regimes, to better account for contemporary military technologies.

In Chap. 9, we draw together the threads of this book to explore how these scenarios inform our understanding of Antarctic governance, law and geopolitics.

Reference

Commonwealth of Australia. 2016. *Australian Antarctic Strategy and 20 Year Action Plan.* Canberra: Commonwealth of Australia. https://www.antarctica.gov.au/about-us/antarctic-strategy-and-action-plan/. Accessed 28 July 2021.

Chapter 9
Scenario Analysis and the Classical View of Antarctic Geopolitics

Abstract In this chapter, we explore how the scenarios constructed in Chap. 8 help us think about Antarctic geopolitics, conceived as the intersection between power, authority, and space, in rivalry between states. We suggest that viewing Antarctic geopolitics through this lens can assist in closing some important conceptual gaps in international law and institutionalist perspectives on Antarctic governance. We examine the intersections between spatial ordering, authority, and power in the subtle rivalry between states in Antarctica and the Southern Ocean. Our scenarios illustrate how projection of power may influence the future of the ATS—both globally, from great power competition, and regionally, from simmering rivalries in the Indo-Pacific and South Atlantic. We suggest that the concept of balance of power might usefully be viewed not simply as an adversarial relationship, but also in associational terms, with its potential for founding a just equilibrium between states. In turn, this just equilibrium can enliven the authority of enabling institutions (such as the ATS) and international law. We also highlight how spatial ordering of areas in, below and above Antarctica is subject to significant tensions due to ambiguities in interaction between the Law of the Sea Convention, the Antarctic Treaty, and treaties governing outer space.

Keywords Power in Antarctic geopolitics · Authority in Antarctic geopolitics · Space in Antarctic geopolitics · State rivalry in Antarctica · Balance of power in Antarctica · Associational balance of power · Just equilibrium between states · Spatial ordering in Antarctica

9.1 Introduction

In Chap. 1, we noted that the field of geopolitics encompasses a variety of research approaches, ranging from the classical to the critical, and that these are largely without settled definitions. The definition of geopolitics we have adopted for this analysis is *the intersection between power*, *authority*, and *space in rivalry between states*. We suggested that this understanding of geopolitics suits an analysis of Antarctica—as a region without the usual well-defined practices of territorial sovereignty, and as a place marked by the absence of overt economic and military competition. We

J. McGee et al., *The Future of Antarctica*, Springer Polar Sciences,
https://doi.org/10.1007/978-981-16-7095-4_9

suggested that these factors make 'rivalry' between states a more subtle affair in Antarctica, hence our proposal for a fresh definition.

In this chapter, we explore how the scenarios constructed in Chap. 8 might help us think about the future of Antarctic geopolitics, as we have conceived it. We suggest that viewing Antarctic geopolitics through this lens might assist in closing some important conceptual gaps in international law and institutionalist perspectives on Antarctic governance.

9.2 Geopolitics: The Intersection of Power, Authority, and Space

In this section, we unpack our constituent elements of geopolitics: *power, authority, and space, in rivalry between states*. We explore how the combination of these elements offers a useful approach to assessing the subtle forms of rivalry between states that occurs within the Antarctic region.

9.2.1 Power

There is no shared concept of 'power' within the social sciences. Instead, there is a voluminous literature, replete with different understandings of its meaning and nature. Analysis is conducted at various levels, including social organisation, epistemological approaches, and disciplinary perspectives. Prominent examples include Mills (1970), Dahl (1961), Bachrach and Baratz (1962), Morgenthau (1967), Foucault (2002), Mearsheimer (2001) and Till (2018). We do not revisit these debates here; rather, we adopt a modest and pragmatic approach of settling upon a broadly accepted understanding of the concept of power, that is useful in our analysis. However, we acknowledge that many of these other conceptions have their strengths.

In our view, the work of Lukes (2005) provides a clear and pragmatic lens. In his seminal work, Lukes identifies 'three faces power', which can be thought of as the ability of an actor to:

- *influence the decision-making of another actor* (the first face);
- *set the agenda or control the parameters of discussion* which leads to the decision-making of another actor (the second face); and
- *manipulate and control what another actor is able to think* in relation to an issue (the third face).

Lukes' work illustrates three broad, but distinct, understandings of how power can be applied to human social relations—in any context and at any scale, including international relations between states.

In keeping with our broad orientation towards classical geopolitics, our primary unit of analysis is the state and its interaction with other states in Antarctica and adjoining regions. Here, we draw primarily upon Lukes' first and second faces of power (listed above). States frame arguments to further their interests in regard to legal and other obligations within the international system. This framing may well influence the agenda or control the parameters of discussion and decision-making within international forums. However, in the context of Antarctic geopolitics, we find it difficult to see how a corporate entity such as a state might be subject to ideological control from another state, with manipulation and command over what 'it', as a corporate entity, thinks. We therefore see Lukes' third face of power as less applicable to this analysis.

Having outlined our understanding of 'power', we will consider how Lukes' first face of power plays out in the militarisation scenarios developed in Chap. 8.

9.2.2 Military Power

Perhaps the most obvious way a state might influence another's decision-making is through actual, or threatened, use of military force. However, a key aim of the Antarctic Treaty is to maintain a non-militarised region in the area below 60° South. For this reason, the military capacities of states (either within the Antarctic region, in adjoining regions, or at a global level) have not, at least publicly, been a significant factor when assessing the ability of one state to influence the decision-making of another in relation to matters concerning Antarctica and the Southern Ocean. For instance, writing in the late-1980s, Joyner (1989) pointed out that, since the late 1950s, the non-militarisation provisions of the Antarctic Treaty had effectively created a military *balance of power* in the Antarctic region. Within the Antarctic Treaty Area (ATA), no country has a significant advantage over any other in terms of projecting hard military power; that is, the presence and operation of naval ships, military aircraft and troops. This is an important observation by Joyner. Nonetheless, military forces are present routinely in the ATA. They provide a vital role in support of peaceful, civilian activities, such as supply of Antarctic research stations and assisting with the logistics of scientific research.

The great powers, such as the United States, China and the Russian Federation, have naval forces capable of projecting naval power over long distances, extending to the northern expanses of the Southern Ocean. The near Antarctic states—Argentina, Australia, Chile, New Zealand and South Africa possess a significant geographic advantage that allows their navies (where they have icebreaking, or ice-strengthened capabilities) to operate within the ATA. At present, the New Zealand Defence Force is considering adding an ice-strengthened, offshore patrol vessel to its fleet (Fish 2021).

Notwithstanding the actual and potential presence of military forces in Antarctica and the Southern Ocean, it can be argued that Joyner's (late-1980s) thesis about a military balance of power in the Antarctic region still holds true today. States have relatively limited ability (or incentive) to use traditional military means to influence the decision-making of other states operating in the region, *when compared to other areas of the planet*. There are several reasons for this. First, the non-militarisation provisions of the Antarctic Treaty place stringent restrictions on the use of military forces within the ATA. Second, there is the relative absence of strategic assets (or targets) in the region, such as cities, populations and readily accessible terrestrial resources. Third, there is the immense geographical remoteness and severe climatic conditions of the region, which hinder military access and operations within the ATA. Thus, Joyner's balance of power thesis seems useful when contemplating the effect of Article IV of the Antarctic Treaty.

However, the scenarios developed in Chap. 8 challenge the notion that the projection of military power in Antarctica is attained primarily through boots on the ground, ships on the sea, and planes in the air. Given developments in military and space technologies, a more contemporary analysis is needed. As we saw in Chap. 6, in recent years there have been significant technological advances in unmanned aircraft, unmanned naval vessels, autonomous vehicles and vessels, ocean sensing systems and satellite technologies. This has amplified the power projection capabilities of near Antarctic states; of states that are less proximate, but technologically advanced; and of the great powers—wherever they are located. More remote, less technologically advanced, and smaller states are increasingly at a disadvantage when it comes to the projection of military power within Antarctica and the Southern Ocean. Our scenarios illustrate the potential for short-term, seamless re-purposing of civilian equipment located within the ATA towards military purposes—particularly within a wider network of connected command, control, communications and reconnaissance (CCCR) systems.

As we have seen, the ATA has remained non-militarised over the past 60 years in terms of traditional modes of military activity. There are significant material and institutional barriers to the overt use of military forces for non-peaceful purposes. However, both the *Stealth* and *Cold War II* scenarios suggest that technological developments increase the strategic advantage of conducting some types of military activity within the region. This raises the prospect of a more conflictual future for Antarctica, in which Lukes' first face of power may become more prominent in relations between states.

We now turn to Lukes' second face of power—the ability to set the agenda, or control the parameters of discussion, which leads to the decision-making of other actors. For this, we shift our gaze to the projection of power through Antarctic science and logistics.

9.2.3 Scientific and Logistical Power

Power in Antarctica is also demonstrated by states' scientific and logistical capabilities on the Antarctic continent and in the Southern Ocean. As we noted in Chap. 4, Quilty refers to 'science as the currency of credibility' in Antarctica (Quilty 1990).

We think it goes further. Clausewitz famously said that, 'War is not merely a political act, but also a real political instrument, a continuation of political commerce, a carrying out of the same by other means' (Clausewitz 1918). In Antarctica, we might say that *science* is the continuation of politics by other means. This is not to detract from the value of Antarctic science *as science*. It is an observation of the dual role that science plays in the area below 60° South.

Scientific and logistical capabilities in Antarctica are prominent markers of diplomatic leadership and influence. Strong competencies in these areas (particularly in *both* areas) enables states to exert greater influence over what issues are introduced and which topics receive prominence within the decision-making bodies of the Antarctic Treaty System (ATS) (Haward and Griffiths 2011). This is not limited to scientific matters. Within the Antarctic Treaty Consultative Meetings (ATCM), Committee for Environmental Protection (CEP) and *Commission for the Conservation of Antarctic Marine Living Resources* (CCAMLR) meetings, those states with well-established scientific and logistical capacities wield greater diplomatic influence and capacity to set agendas across a wide range of issues, including the non-militarisation and inspection aspects of the Antarctic Treaty. As Dudeney and Walton (Dudeney and Walton 2012) found, the original signatories to the Antarctic Treaty were dominant in publishing scientific papers on Antarctica, but also in submitting Working Papers to the ATCM, which is the institutional mechanism for the ATCM to make legally binding decisions.

How do we gauge this type of 'power'? At present, there are no comprehensive scorecards or rankings available of states' scientific/logistical influence in Antarctica. There is no readily available and transparent means of comparing the respective financial commitments of states to their national Antarctic programs. The proxy measures that might be relied on to identify scientific/logistical capabilities are data such as the number of year-round research stations and quantity of scientific publications. For stations, the latest station catalogue produced by the Council of Managers of National Antarctic Programs (COMNAP) (2017) indicates that, as of 2017, there were 76 Antarctic research bases, of which 40 were classified as 'year-round facilities'. The 40 year-round facilities were operated by the following countries: Argentina (6), Russia (5), Australia (3), United States (3), Chile (3), United Kingdom (2), China (2), France (2), India (2), Republic of Korea (2), Japan (1), Brazil (1), Norway (1), Poland (1), South Africa (1), Ukraine (1) and Uruguay (1). Based on these figures, the original 12 signatory states to the Antarctic Treaty operated 65 per cent of the year-round research stations. This does not take into account the relative size of the research stations (the US base at McMurdo is the largest); however, it is still useful as a rough indicator of the ongoing influence of the original signatories.

Similarly, in exploring the extent of authorship of scientific publications on Antarctic science as a proxy measure of influence, Dudeney and Walton (2012) found that, during the period 1992–2010, the seven claimants, plus the US and Russia dominated scientific publications and Working Papers submitted to the ATCM. Similar analysis from Dastidar (2007) for the period 1980–2004 indicated that the original 12 signatories to the Antarctic Treaty authored 72 per cent of scientific publications on Antarctica. Countries such as China and India are building their Antarctic research programs. However, at present, there remains a hierarchy in Antarctic scientific and logistical capabilities that is heavily weighted towards the claimant states and some other signatories—particularly the US and Russia. In terms of projecting scientific and logistical power, then, this group of states tends to exert the most influence. The 'weight' of a state's scientific activity is often associated with its infrastructure and logistical commitments. These can include multiple, year-round scientific research stations, shipping and air access to the continent, and active collaboration with other parties.

Two of our scenarios, *Splendid Isolation* and *Quarantine*, explore the second face of power by illustrating how states that lead in Antarctic science and logistics might set a positive agenda on the issue of Antarctic non-militarisation. In *Splendid Isolation,* these states secure agreement to a new binding measure for transparency on the use of data from telescopes and sensors within the ATA. In *Quarantine,* they propose new rules within CCAMLR to ensure appropriate use of aerial surveillance.

9.2.4 Authority

'Authority' is another concept subject to ongoing definitional debate. Again, it is not our purpose here to revisit these debates or to resolve them. However, for the purposes of this analysis, we adopt the well known Weberian institutional notion of authority that is centred on *legitimacy*. This is articulated by Hird (2007, pp. 60–61) as follows:

> Authority describes relation that exists between an actor and an institution when the actor perceives the institution to be legitimate. Where an actor sees a rule as legitimate, the rule has been internalized into interests and takes on the quality of becoming authoritative over the actor. It is hierarchically superior to the actor and partly determines the actor's behaviour. Two effects of this for compliance with rules are that the rule becomes naturalised and only occasional deviance needs to be policed.

Authority can therefore be understood as a quality of institutions and rules which leads to *compliance*. As a constituent element of our understanding of geopolitics, incorporating a notion of authority broadens our gaze from a simple view of the actions of states in exercising control over the decision-making of other states to a more nuanced understanding of the legitimacy of institutions—and a sense that the rules of those institutions have been internalised by actors and are (at least partly) determinative of their behaviour.

And so, we find that authority is concerned with the role of institutions in ordering a social domain, by creating rules which take on the quality of being authoritative over actors within that domain (Hird 2007). Arguably the leading English legal theorist of the twentieth century, H. L. A. Hart, refers to this phenomenon as the internal point of view (Hart 1961). This manifests as a *sense of obligation*, which social actors owe to the social and legal rules of the society in which they are located. This sense of obligation comes from a subjective belief in the inherent 'rightness' of the social and legal rules of the legal system. As Pettit (2019) argues, this subjective belief in the rightness of social and legal rules stems from an actor's *membership* of a social group, a key element of which is acceptance of the group's social and legal rules.

In the international society of states (Bull 1977), this sense of obligation helps explain how the rules of international law influence the behaviour of states towards compliance, even in situations where compliance may not be in a state's immediate self-interest. In explaining the operation of international law, Franck (1990) describes this as the 'property of a rule or rule making institution which exerts a pull towards compliance on those addressed normatively'. Similarly, Brunée and Toope (2010) identify that, under certain conditions, states manifest a sense of obligation to the rules of international law that can pull them towards compliance. Our understanding of authority therefore focuses on *the role of the legitimacy of institutions and rules in shaping the behaviour of states*. In the specific context of Antarctic geopolitics, we suggest that 'authority' refers to the ability of key regimes, such as the ATS, to create rules which states believe have legitimacy, in the sense of an inherent obligation with which they should comply.

Our understanding of authority, arising through the legitimacy of rules and the inherent sense of obligation states have to such rules, is quite separate from the concept of power discussed above. It accepts that international institutions and rules are created by states with varying interests. However, such institutions and rules may develop sufficient legitimacy to be internalised by the states, with a resultant strong sense of subjective obligation to comply with such rules. However, rules and institutions with legitimacy can also become 'sticky', in the sense that, once created, states find it hard to depart from them.

All five of our scenarios have authority as a background concept to the operation of the ATS. Each scenario assumes that the ATS has a level of authority (through the legitimacy of its rules and institutions) to at least partly shape state behaviour through a sense of obligation to the rules of the system. Some of the scenarios, particularly *Stealth* and *Cold War II*, explore plausible futures where the regime's rules of non-militarisation come under significant strain and among key states the sense of obligation to such rules starts to fray. Regardless of whether authority is maintained or it unravels, the concept is a key element of our understanding of Antarctic geopolitics.

9.2.5 Space

In line with leading classical understandings of geopolitics, we use the term 'space' to refer to the geographic components of the physical and material environment which affect international behaviour. Spatial characteristics and differences are two key influences upon the behaviour of states in international systems. As Starr (2013, p. 435) comments:

> The spatial and geographic components of the environment of any international actor are thus essential to understanding choice in foreign policy and international relations. The geopolitical environment includes the effects of space, topography, position, and climate (Osterud

> 1988). Harold (1963) has asserted that geopolitical hypotheses deal with the configuration and layout of lands and seas, climate, and the distribution of natural resources, as well as those dealing with distribution of people, social institutions, or behavioural patterns. Thus, all international actors or entities may be located spatially - in some geopolitical arrangement of topography and some distribution of people behaviour and resources.

Similarly, Kelly (2019, p. 48) observes:

> Space forms the crux of geopolitics - comparative land and sea locations, positions of states, regions, and resources, and a multiple of other territorial features - distance and isolation, climate, topography and migrations, maritime and land orientations, areas of pivot and leverage whether core or periphery, and so forth.

Aside from the importance of spatial factors on state behaviour, space is also significant for the way that institutions and rules order the planet's land, oceans and airspace. This was noted by the (controversial) German international lawyer Schmitt (2006), who, despite coming from a continental juridical background, draws heavily on the British theorist of classical geopolitics, Mackinder (1904). Schmitt describes what he terms the 'Nomos of the Earth'—the 'immediate form in which the political and social order of a people becomes spatially visible' (2006, p. 70). Schmitt notes the historical development of a spatial ordering of all the land and oceans of the Earth in European international law. For our purposes, the concept of 'space' encompasses both the *geographic characteristics* of the physical environment that impact state behaviour and the *forms of spatial ordering* (through institutions and rules) constructed to facilitate social and political order between states.

Thus, our conception of classical geopolitics focuses on the intersection of *power, authority, and space in their role in rivalry between states*. As we have seen, this rivalry takes many forms. The most obvious is military competition, though we have argued that this is less relevant within the ATA. Rivalry may also manifest as competition for influence over places or resources, as market share in export markets, or as control over new technologies. Rivalry may occur in the social and cultural interactions between states, and in states' efforts to frame past and current events in the service of their strategic and/or economic interests.

9.3 Antarctica Through a Classical Geopolitical Lens

So far, we have explored three concepts that, we suggest, are in combination useful for analysing the various forms of rivalry between states over Antarctica: power, authority, and space. Next, we therefore examine how the scenarios developed in Chap. 8 might inform our thinking about the *intersections* between these three elements.

Our five scenario narratives were constructed from various combinations of the top two critical uncertainties for Antarctic militarisation identified in Chap. 7. One of these is the *level of stability of the international system*. Primarily, this concerns Lukes' first face of power—the ability of a state to use its military and economic resources to influence the decision-making of other states.

By engaging with an *exogenous* variable (stability of the international system), these scenarios adopt a *macroscopic* view (see Chap. 4) which is wider than many of analyses of Antarctic governance and futures we have encountered. These analyses tend to focus primarily on one region (Antarctica), and on one international regime (the ATS). We query the contemporary relevance of a traditional view of the 'separateness' or 'exceptionalism' of the ATS in managing the region, with its limited account of the spillover impacts of wider issues and their institutions, such as the United Nations. Our scenarios assume the projection of power between global players in the broader international system (and the level of tension this produces) is pivotal to the future of non-militarisation in Antarctica.

A key uncertainty in these scenarios (particularly *Quarantine* and *Cold War II*) is the extent to which competition between the great powers might spill into the Antarctic region and the ATS. A large slice of the scholarship on Antarctic governance notes that, to date the ATS has been able to insulate itself from wider international tensions. As noted in Chap. 2, a key example is the formation of the Antarctic Treaty 1959 during a period of heightened superpower tension. Our scenarios challenge the assumption that rising tensions between the great powers will be quarantined to areas north of the ATA. Of course, we are not alone in raising this concern.

Importantly, these scenarios explore how the 'balance of power' concept might play out in Antarctica. Here, we see a crucial intersection between the concepts of power and authority. As mentioned above, balance of power theory suggests that a fundamental part of states' ability to successfully navigate differences in levels of power is to form *balancing coalitions* of weaker states. These coalitions can serve as a significant deterrent against a powerful state deploying military force against a weaker one. However, as Kelly (2019) and Sloan (2017) point out, 'realist' international relations theory (from which the concept of the balance of power is drawn) tends to focus on the aggregate military and/or economic power of states. A more nuanced analysis might assess the power differentials between states at a regional or sub-regional level. Kelly and Sloan both argue that geopolitics (in emphasising spatial differences and power relationships) offers a more fine-grained approach to analysing the respective power relations between states, and to assessing the power imbalances in a given geographical context. We agree that a geopolitical lens is useful for analysing the balance of power at a regional and sub-regional level.

For this reason, our scenarios probe whether power balancing efforts within the wider Indo-Pacific region (such as the 'Quadrilateral' relationship between the US, Japan, India and Australia) might stabilise great power competition within the region that can otherwise spill into Antarctica. Similarly, the scenarios explore whether great power competition between Russia and the European Union (EU) might, in future, impact the ATS. While limited in terms of its participants and iterations, the scenario building process illustrated in Chaps. 7 and 8 suggests the balance of power concept is a key lever for states wishing to successfully navigate great power competition in the broader international system and lessen the prospects of its leakage into Antarctica. We suggest this point is under-emphasised in the literature on Antarctic geopolitics and governance.

9.3.1 Power and International Law

The concepts of great power competition and the balance of power may also be useful for theorists of Antarctic law. The relationship between power and law within international legal scholarship is one of some ambivalence and confusion. It is generally approached through one of three lenses. First, *traditional* or *doctrinal* legal scholarship is focused on explicating the structure and content of the rules of international law. However, it does not theorise much about the social forces (such as power) within the international system that create and influence this law (Cassese 2005). Second, and in contrast, *critical* international law scholarship views the power of key states and/or class conflict as directly shaping the content and outcomes of international law see, for example, Marks (2008), Bowring (2008), and the leading post-structuralist critiques of Kennedy (1987) and Koskenniemi (2005). The critical approach thus engages closely with the concept of power and its influence upon the law, but mainly as a point of critique or deconstruction. Third, the *rational choice* approach to international law, particularly dominant in US legal scholarship (Goldsmith and Posner 2005), applies micro-economic assumptions to argue that state interests are key drivers in the formation of, and compliance with, international law. This view can lead to an overly simple conclusion that powerful states only comply with international law when, on weighing the costs and benefits, it is in their interests to do so.[1]

Within international relations scholarship (and particularly with 'realist' authors), the rational choice approach has led to a focus on adversarial theories of the balance of power. Primarily, these theories focus on how to secure an equilibrium in military and economic power between the key states in a region, to check the expansionist desires of a rising state—see Mearsheimer and Walt (2016). If we were to embrace Mearsheimer's stance that international law and institutions are (at best) marginally relevant to the behaviour of states (Mearsheimer 2001), then the possibilities for international law in this context are bleak. On this view, potential conflict is managed only through the self-interested calculations of states that flow from the threat of military action. We suggest that this view is unduly dismissive of the role that international law can play.

However, there is a fourth (and sometimes overlooked) perspective on the relationship between power and international law. This approach also focuses on the balance of power, but is concerned with its *associational* characteristics (McGee et al. 2017). By this, we mean striking a functional balance of military and economic power between key states in a region that prevents one state from dominating the practices of international rule-making. The importance of an *associational balance of power* was noted by the leading (early twentieth-century) international lawyer, Lassa Oppenheim, who described it in the first edition of his classic 1905 textbook as a pre-condition to law (1905, p. 73).

---

[1] This approach often draws on the 'structural realism' of Kenneth Waltz, which is based on rational choice analysis (Waltz 2010).

> The first and principal moral that can be deduced from the history of development of the law of nations is that a law of nations can only exist if there be an equilibrium, a balance of power, between the members of the family of nations.

This notion was drawn on by Bull, who commented that international law 'depends for its very existence as a system of rules on the balance of power' (2012, p. 104). As he explains:

> Where one state is preponderant, it may have the option of disregarding the rights of other states, without fear that those states will reciprocate by disregarding their rights in return. It is this feeling that there must be some security for the observance of rules of international law, other than mere hope that a preponderant state will choose to be law abiding, that leads international lawyers like Oppenheim to the conclusion (above) (2012, p. 104).

More recently, the associational balance of power has featured in the work of international relations theorist, Richard Little. Little suggests that the concept was extended from domestic contexts to the international system, as follows:

> It was acknowledged that power need not be used coercively in the international arena to promote the narrow sectional interests of individual states, but communally to sustain a just equilibrium which would reflect the interests of all the members of the system (1989, p. 95).

Importantly, an associational balance of power is not an inevitable or natural feature of the international system (Little 1989). As we find with the establishment of a mixed government in a domestic setting, associational balances of power need to be created, re-created, and protected by the ongoing, self-conscious effort of states. As Little comments, 'to operate, there must be in existence a "just equilibrium" arrived at by mutual agreement' (1989, p. 95). This idea of a *just equilibrium of power* within a region invites us to re-engage with the marginalised concept of the balance of power. Our scenarios suggest that this may be important to the future wellbeing of Antarctic law and governance.

The associational balance of power of Oppenheim, Bull and Little is very different from *adversarial* versions of the balance of power. The 'balancing' of military and economic strength in a region is not for the purpose of generating fear of military reprisal in an expansionist state, but rather to allow leading states the capacity to develop the conditions that will allow for a just and mutual power sharing through the operation of the rules of international law. With an associational balance of power, a rough balancing of military and economic power within a region provides conditions in which rules formed can represent all states' interests, rather than being dictated by a hegemon. This provides the conditions of reciprocity between states, and builds legitimacy—the sense of legal obligation that can pull states towards compliance, regardless of their self-interested calculations (Brunée and Toope 2010). We suggest that striving for an associational balance of power at a regional level is important if we are to create the baseline conditions where the rules of international law can be strengthened through legitimacy and a sense of legal obligation (Hird 2007).

The associational balance of power lies at the intersection between two of our core elements of geopolitics: power and authority. It is a more optimistic and expansive view of the potential role of international law in Antarctic affairs.

Our scenarios suggest that finding a just equilibrium within adjoining regions (such as within the Indo-Pacific or between Russia and the EU) has potential to stabilise great power relations. It has potential to strengthen the role of international law and to prevent great power competition spilling into the ATA. For this reason, we suggest that close attention be paid to assessing balance of power relations both within the ATA *and* in regions of tension that adjoin it, such as the Indo-Pacific and South Atlantic. As noted above, following Joyner (1989), we might find a good example of a just equilibrium of power in the non-militarisation of the ATA. Indeed, we suggest, this has been foundational to the high level of legitimacy the ATS has enjoyed for the last 60 years (Young 2010).

The value of classical geopolitical scenario analysis to pursuing a just equilibrium is that it can map plausible futures at a regional level within Antarctica, including linkages with adjoining regions and the wider international system. This can assist policy makers to assess plausible pathways to an associational balance of power, and to build the case for institutional capacities and adaptiveness to maintain a just equilibrium. In turn, this helps maintain the authority of the international law that underpins the ATS. For these reasons, our scenarios suggest that the links between power and authority in Antarctic governance should be of interest to international lawyers and to theorists of international relations.

9.3.2 Antarctica and 'Space in Rivalry Between States'

Our other top critical uncertainty for Antarctic militarisation (see Chap. 7) was the level of military/strategic advantage from the use of Antarctica and the Southern Ocean for non-peaceful activities, such as military research and substantive military activity. As noted in Chap. 7, writing in the late 1980s, Parsons (1987) made the astute observation that the Antarctic Treaty is a special case because, in that unlike in the case most arms control agreements, the strategic value of military activity in the ATA had, since the Treaty's inception, been static or diminishing.

As we have seen, there has been no obvious strategic reason for states to deploy troops, ships and aircraft to the Antarctic continent, or into adjacent areas of the Southern Ocean. With the exception of the Falklands/Malvinas war, for the past 60 years, Antarctica has been far removed (spatially) from key areas of military engagement across the globe. The ATA also enjoys a significant geographic advantage in being remote from viable points of deployment to active theatres of military conflict. Parsons' point is that demand for use of the ATA for military activity was comparatively weak (as at the late 1980s) because there are better places to carry this out.

However, we may have turned a corner. Our scenarios suggest plausible futures in which the strategic advantage of Antarctic militarisation increases. As we saw in Chaps 6–8, rapid advances in technology may escalate the strategic value of Antarctica and the Southern Ocean as part of a global network for the receipt and transmission of information of potential military use. This is interesting in terms of the spatial element of geopolitics. The spatial ordering of Antarctica that is provided by the Antarctic Treaty applies to areas of ocean, ice and land in the area below 60° South. In the ocean, the Treaty applies to the water surface, water column, sea floor and subsoil areas. This is subject to high seas freedoms reserved under Article VI of the Antarctic Treaty and possible application of Part XI of the United Nations Convention on the Law of the Sea (UNCLOS) to mining activities on the ocean floor. Above the surface of the ocean and ice, the Antarctic Treaty also applies to airspace up to the (most accepted, but problematic) legal boundary between airspace and outer space, at 100 km above the Earth's surface—the Kármán Line (Nelson 2019; Fernández and Córdoba 2004; Sanz Fernández 2004).

The spatial ordering provided by the ATS leaves open the possibility of overflight by satellites at altitudes above 100 km. This is likely to fall outside the jurisdiction and control of the Antarctic Treaty parties. Since the 1960s, the key space powers have launched numerous polar-orbiting satellites which circle the Earth approximately every 60 min, with potential for both civilian and military uses, such as weather forecasting, communications and surveillance. From the very beginning of the 1959 Antarctic Treaty, there was a potential weakness in its non-militarisation provisions: satellites could pass right above the ATA and be engaged in activities of a military nature. Our scenarios (particularly *Quarantine* and *Pressures Managed*) suggest that, over the next decade, the current delineation between Antarctic airspace and outer space, and the vacuum in Antarctic governance above the Kármán Line, may come under increasing pressure.

As we saw in Chap. 6, Antarctica is one of the best places on the planet for astronomical research: the continent rises to an elevation of over 4 km and is devoid of ambient light for large parts of the year. These conditions provide ideal opportunities for astronomical research using radio telescopes, infrared telescopes and associated ground receiving stations. Astronomical research infrastructure is being actively developed by several key Antarctic states, including the US, Germany and China. As we saw in Chaps. 7 and 8, technological advances in satellite and military communications in the US, China, Russia and some European countries are opening up prospects for the development of globally connected command, control, communications and reconnaissance (CCCR) systems. These rely on satellites to uplink and downlink information with land- and ocean-based sensors and transmitters. These CCCR systems may also increase demand for use of Antarctic-based radio telescopes, ground station receivers and ocean sensors for use in such military communication systems. Similarly, it is reasonably foreseeable that infrared telescopes located in Antarctica may be called on for surveillance and tracking of long-range missiles and other ordinance systems.

Our scenarios illustrate several plausible (and disturbing) futures for deployment of these technologies, both in areas above the ATA and in waters immediately around Antarctica, by states asserting high seas freedoms. These scenarios highlight difficulties with the current spatial ordering of the area above Antarctica, and the challenges for the ATS in regulating the interface between space-based assets and scientific infrastructure located in Antarctica, which may be repurposed seamlessly to interface with military systems. This 'dual-use' problem has been raised by several Antarctic states, but to date there has been no significant movement by Treaty parties to address this (admittedly sensitive) issue. We suggest that, for those Parties seeking to maintain the non-militarisation of Antarctica, continuing to look the other way may not be a viable long-term strategy.

Our scenarios also point to difficulties in the spatial ordering of the ATS in its intersections with UNCLOS. As noted above, the Antarctic Treaty applies below 60° South, and the CAMLR Convention applies to sections of the Southern Ocean that extend further north to the Antarctic convergence. However, there is significant ambiguity and differences of interpretation between key Antarctic states over application of the Antarctic Treaty to marine areas outside the icesheets and land sections of the continent. As touched on in our *Cold War II* scenario, authors from both the US and China have published arguments suggesting that the high seas freedoms preserved in Article VI of the Antarctic Treaty effectively mean that 'non-aggressive' military activity (that is, activity not in breach of the general obligations of non-aggression under the United Nations Charter) might be carried out legally within Antarctica's oceans—including waters adjacent to the Antarctic coast.

To provoke thinking on this point, the scenarios *Stealth* and *Cold War II* draw attention to the ambiguous spatial ordering of high seas freedoms at the interface between UNCLOS and the Antarctic Treaty. This leaves open the possibility of a future in which military training or other 'non-aggressive' military activity is carried out in the oceans below 60° South. Clearly, this would mark a significant break from current practice and the declared values of the Treaty parties.

9.4 Summary

This chapter has explored how scenarios on Antarctic militarisation might inform a classical geopolitical understanding of Antarctica and the ATS. We have unpacked the constituent elements of our conception of geopolitics—as power, authority and space in rivalry between states and examined their connections.

In particular, these scenarios illustrate how projection of power may influence the future of the ATS—both globally, from great power competition, and regionally, from simmering rivalries in the Indo-Pacific and South Atlantic. We suggested that we might view the balance of power not simply as an adversarial relationship, but also as an associational balance of power, with its potential for a just equilibrium between states, which in turn enlivens the authority of enabling institutions and international law. We suggest that understanding the linkages between power and authority is particularly important for anticipating plausible futures for Antarctic governance.

Finally, we explored the third element of our conception of classical geopolitics: space—in this case the spatial ordering of the ATS. We saw that the spatial ordering of areas in, below and above Antarctica is subject to significant tension due to ambiguities in interactions between UNCLOS, the Antarctic Treaty, and treaties regulating outer space. This plays out in several of our scenarios, where some states exploit these ambiguities to deploy space- and ocean-based assets for military CCCR systems. As we have seen, these risks are compounded by difficulties in regulating and monitoring dual-use technologies within the ATA.

In this chapter, we have explored the intersections between spatial ordering, authority and power in the subtle rivalry between states over Antarctica. In Chap. 10, we conclude this volume with a discussion of our research questions, key findings and directions for future work.

References

Bachrach, Peter, and Morton Baratz. 1962. Two Faces of Power. *American Political Science Review* 56 (4): 947–952.

Bowring, Bill. 2008. *The Degradation of the International Legal Order? The Rehabilitation of Law and the Possibility of Politics*. London: Routledge-Cavendish.

Brunée, Jutta, and Stephen J. Toope. 2010. *Legitimacy and Legality in International Law: An Interactional Account*. Cambridge: Cambridge University Press.

Bull, Hedley. 1977. *The Anarchical Society: A Study of Order in World Politics*. New York: Columbia University Press.

Bull, Hedley. 2012. *The Anarchical Society: A Study of Order in World Politics*, 4th ed. London: Red Globe Press.

Cassese, Antonio. 2005. *International Law*, 2nd ed. Oxford: Oxford University Press.

Clausewitz, Carl von. 1918. *On War* (trans. Graham, J. J.) London: Kegan Paul, Trench, Trubner & Co. https://oll.libertyfund.org/page/clausewitz-war-as-politics-by-other-means. Accessed 19 August 2021.

COMNAP. 2017. *Antarctic Station Catalogue*. Christchurch: Council of Managers of National Antarctic Programs. https://www.comnap.aq/publications/. Accessed 19 August 2021.

Dahl, Robert A. 1961. *Who Governs: Democracy and Power in an American City*. New Haven: Yale University Press.

Dastidar, Prabir G. 2007. National and Institutional Productivity and Collaboration in Antarctic Science: An Analysis of 25 Years of Journal Publications (1980–2004). *Polar Research* 26 (2): 175–180.

Dudeney, John R., and David W. H. Walton. 2012. Leadership in Politics and Science within the Antarctic Treaty. *Polar Research* 31 (1): 11075.

Fish, Tim. 2021. New Zealand releases Southern Ocean patrol ship RFI. *Janes*. Defence News, 17 May. Surrey, UK: Jane's Group UK Limited. www.janes.com/defence-news/news-detail/new-zealand-releases-southern-ocean-patrol-ship-rfi. Accessed 4 June 2021.

Foucault, Michel. 2002. *Power*. Ed. James D. Faubion, London: Penguin.

Franck, Thomas M. 1990. *The Power of Legitimacy Among Nations*. New York: Oxford University Press.

Goldsmith, Jack L., and Eric A. Posner. 2005. *The Limits of International Law* Oxford: Oxford University Press.

Hart, H.L.A. 1961. *The Concept of Law*. Oxford: Oxford University Press.

Haward, Marcus, and Tom Griffiths, eds. 2011. *Australia and the Antarctic Treaty System: Fifty Years of Influence*. Sydney: UNSW Press.

Hird, Ian. 2007. *After Anarchy: Legitimacy and Power in the United Nations Security Council*. Princeton and London: Princeton University Press.

Joyner, Christopher C. 1989. Non Militarization of the Antarctica: The Interplay of Law and Geopolitics. *Naval War College Review* 42 (4): 83–104.

Kelly, Phil. 2019. Rescuing Classical Geopolitics: Separating Geopolitics from Realism. *Geopolitics, History and International Relations* 11 (1): 41–58.

Kennedy, David. 1987. *International Legal Structures*. Baden-Baden: Nomos Verlagsgesellschaft.
Koskenniemi, Martti. 2005. *From Apology to Utopia: The Structure of International Legal Argument*. Cambridge: Cambridge University Press.
Little, Richard. 1989. Deconstructing the Balance of Power: Two Traditions of Thought. *Review of International Studies* 15 (2): 87–100.
Sanz Fernández de Córdoba, S. 2004. 100km Altitude Boundary for Astronautics. FAI Astronautic Records Commission (ICARE). Lausanne: Fédération Aéronautique Internationale. https://www.fai.org/page/icare-boundary. Accessed 19 August 2021.
Lukes, Steven. 2005. *Power: A Radical View*, 2nd ed. Basingstoke, UK: Palgrave Macmillan.
Mackinder, Halford J. 1904. The Geographical Pivot of History. *The Geographical Journal* 23 (4): 421–437.
Marks, Susan. 2008. Introduction. In *International Law from the Left: Re-examining Marxist Legacies*, ed. Susan Marks, 1–29. Cambridge: Cambridge University Press.
McGee, Jeffrey, Brendan Gogarty, and Danielle Smith. 2017. Associational Balance of Power and the Possibilities for International Law in the South China Sea. *Asia-Pacific Journal of Ocean Law and Policy* 2 (1): 88–116.
Mearsheimer, John J. 2001. *The Tragedy of Great Power Politics*. New York: W. W. Norton.
Mearsheimer, John J., and Stephen M. Walt. 2016. The Case for Offshore Balancing. *Foreign Affairs* 95 (4): 70–83.
Mills, C. Wright. 1970. *The Power Elite*. Oxford: Oxford University Press.
Morgenthau, Hans. 1967. *Politics Among Nations: The Struggle for Power and Peace*, 4th ed. New York: Knopf.
Nelson, Timothy G. 2019. Where Does Space Begin? The Decades-Long Legal Mission to Find the Border Between Air and Space. *SpaceNews*, 26 March. Boulder, CO: Pocket Ventures LLC. https://spacenews.com/op-ed-where-does-space-begin-the-decades-long-legal-mission-to-find-the-border-between-air-and-space/. Accessed 19 August 2021.
Oppenheim, Lassa. 1905. *International Law: A Treatise*, vol. 1. London: Longmans Green.
Parsons, Anthony, ed. 1987. *Antarctica: The Next Decade. Report of a Group Study Chaired by Sir Anthony Parsons*. Cambridge: Cambridge University Press.
Pettit, Philip. 2019. Social Norms and the Internal Point of View: An Elaboration of Hart's Genealogy of Law. *Oxford Journal of Legal Studies*. 39 (2): 229–258.
Quilty, Patrick G. 1990. Antarctica as Continent for Science. In *Antarctica's Future: Continuity or Change?* eds. Richard A. Herr, H. Robert Hall and Marcus G. Haward, 29–37. Hobart: Tasmanian Government Printer.
Sanz Fernández de Córdoba, S. 2004. 100km Altitude Boundary for Astronautics. FAI Astronautic Records Commission (ICARE). Lausanne: Fédération Aéronautique Internationale. https://www.fai.org/page/icare-boundary. Accessed 19 August 2021.
Schmitt, Carl. 2006. *The Nomos of the Earth in the International Law of Jus Publicum Europaeum* (trans. Ulmen, G. L.). New York: Telos Press Publishing.
Sloan, Geoffrey. 2017. *Geopolitics, Geography and Strategic History*. London: Routledge.
Starr, Harvey. 2013. On Geopolitics: Spaces and Places. *International Studies Quarterly* 57 (3): 433–439.
Till, Geoffrey. 2018. *Seapower: A Guide to the Twenty-First Century*, 4th ed. London: Routledge.
Waltz, Kenneth N. 2010. *Theory of International Politics*. Long Grove, IL: Waveland Press.
Young, Oran R. 2010. *Institutional Dynamics: Emergent Patterns in International Environmental Governance*. Cambridge MA: MIT Press.

Chapter 10
Conclusion

Abstract The current (and largely unforeseen) COVID-19 pandemic highlights the value of scenario analysis as a complementary exercise to standard, extrapolative prediction. In this chapter, we review our main findings for geopolitical scenario analysis in general, and for Antarctic geopolitical futures in particular. We conclude that the Antarctic Treaty promotes effective governance of a region described in the Madrid Protocol as 'a natural reserve devoted to peace and science'. We hope to have shown that a classical geopolitical lens is important and relevant to the study of Antarctic futures. On the specific topic of militarisation, we identified some key driving forces for change and equilibrium. How well the ATS responds to these driving forces will turn on its resilience as a governance system. By this, we mean 'a capacity to prepare for, to respond to, or bounce back from problems or perturbations and disturbances, which cannot necessarily be predicted or foreseen in advance' (Chandler and Coaffee 2017). As we have seen, scenarios are useful in this zone beyond standard prediction—*provided* they are plausible, rigorous, and robust. It is our hope that like-minded Parties and researchers might collaborate in scenario work, to contribute to the resilience of the ATS in the challenging years ahead.

Keywords Wind-tunnel testing of strategy · Entertainment scenarios · Scenarios and methodological chaos · Bifocal approach to Antarctic sovereignty · Definition of military activity · Antarctic treaty compliance and inspection · Resilience of Antarctic governance

Work on this book began in 2019. By February 2020, the SARS-CoV-2 virus had been named, and the disease it caused (COVID-19) was spreading through human populations across the globe. As we finish this book, over 210 million people have been infected, 4.4 million have died, and more contagious variants threaten the efficacy of current interventions. All nations are in a race to vaccinate their entire populations, and (as usual) the poorest countries suffer most. Immense social and economic impacts continue, and we have, by and large, submitted to curbs on our freedoms unparalleled in peacetime. As we write, more than half the Australian population is in 'lockdown', unable to leave our homes except for the most essential reasons. And yet, in 2019, few of us could have predicted (with confidence) these momentous events. The limits of standard prediction were breached with breathtaking speed.

J. McGee et al., *The Future of Antarctica*, Springer Polar Sciences,
https://doi.org/10.1007/978-981-16-7095-4_10

For us, the current pandemic underscores the value of scenario analysis as a complementary exercise to standard, extrapolative prediction. It also demonstrates that, when developing our scenarios, we must not be too limited in what we consider to be *plausible.* We suspect these lessons apply equally to the future geopolitics of Antarctica.

10.1 Our Objectives

At present, there is growing interest in the future of Antarctic governance. For the last 60 years, the Antarctic Treaty System (ATS) has been comparatively successful in meeting its objectives, and sufficiently adaptive in the face of change. However, within the contemporary literature, there is increasing concern that, in the medium to long term, ATS responses to emerging geopolitical challenges may be inadequate. How do we assess the validity of this concern? How can organisations assess the future robustness of their current strategy? Unfortunately, much of the commentary on Antarctic futures lacks a common language and analytical approach. To address this gap, we set out to explore how the multi-disciplinary focus of geopolitics might be applied *systematically* to Antarctic scenarios.

Clearly, the Antarctic Treaty Area (ATA) is an unusual geopolitical space. This is due to its remoteness, extreme climate, bespoke governance arrangements, unique approach to sovereignty, absence of a permanent population and the fact that most human activity and investment within the ATA emanates from national Antarctic programs. We have suggested that this combination of attributes lends itself to treatment from the primarily state-centric perspective of *classical* geopolitics—a lens that has been underplayed in Antarctic discourse in recent decades. We acknowledge that a classical geopolitics approach will not yield the full story of the driving forces at work in our scenarios. However, we have adopted the classical lens because this perspective yields important insights that are relevant for the central actors of Antarctic governance: nation-states.

To do this, we have drawn on cognate disciplines of geography, strategy, history, international law, and institutional theory. We have attempted to navigate the 'methodological chaos' of scenario theory and practice in a pragmatic way. We have sought to illustrate our findings through a worked example—the emerging issue of Antarctic militarisation. In working through these topics, we have made observations about the theory and practice of geopolitical scenario planning, which could be applied both within and beyond Antarctica. We also sought to examine the uses of classical geopolitical scenarios to inform understandings of the balance of power, including *just equilibrium,* at regional and global scales. Taken together, we hope these various strands will enrich both the theory and practice of scenario planning for Antarctica. Over the rest of this chapter, we review our key findings.

10.2 Our Findings: Geopolitical Scenario Analysis

We suggested that the *limits of prediction* pose a critical risk for any organisation making long-range decisions about its strategy and investments. The accuracy of our forecasts degrades progressively due to the combined effects of *complexity* and *uncertainty*, compounded by *the passage of time*. Scenarios become useful in the zone beyond the limits of standard prediction.

We questioned whether it is *really* the case that scenarios must only start where prediction ends. Based on our experience, we find it hard to believe that most organisations are prepared to divert significant time and resources into scenario planning without some expectation that the scenarios created will provide a credible picture of what the future might be like. For this reason, we proposed that it helps to view scenarios as anticipatory systems, with a distinctive type of predictive model, where prediction is not based on probability, but plausibility. We put forward the idea that, to be useful, scenarios need to operate within a zone of plausibility. We described this as the space in which scenarios provide *evidence-based, broad-brush,* but *targeted prediction*. This is quite different from conventional notions of extrapolative prediction. We contended that other, frequently stated objectives for scenarios—such as creating 'futures consciousness' within an organisation—are necessary, but secondary, to the predictive work that scenarios can provide in the zone of plausibility.

Why is this important for the theory and practice of scenarios? We argued that if we abandon hope that scenarios can provide a limited form of prediction, then we reduce our expectations for methodological rigour, and we fail to harness the potential that a rigorous scenario process can offer. We are left with scenarios that offer little more than what Godet and Roubelat refer to as 'entertainment' (Godet and Roubelat 1996). Of course, we are conscious that a call for scenarios to provide even a limited form of prediction runs counter to much of the scenario literature. For this reason, our hypothesis is tentative, and we welcome further theoretical and empirical research in this space.

From our review of scenario *objectives*, *approaches,* and *techniques*, we suggested that scenario methods can be grouped under one or more of seven broad approaches: *deductive*, *inductive*, *quantitative*, *hybrid*, *integrated*, *scoping,* and *engagement*. Each has its strengths and (at times, considerable) weaknesses. We concluded that no single scenario approach can match the scale and complexity of Antarctic geopolitics. For that reason, we recommend that a scenario process employ a judicious blend of several approaches.

For this book, we focused on one approach, the standard IL 2 × 2 matrix, commonly deployed in geopolitical scenarios. We unpacked it with a worked example and encountered some difficulties applying it to the field of Antarctic geopolitics. We suspect at least some of these difficulties arise when this technique is applied to other fields. This led us to propose some modifications to the IL 2 × 2 technique. We saw that, in the IL 2 × 2 approach, we explore two critical uncertainties (CUs) at their low and high values: CU 1 low, CU 1 high; and CU 2 low, CU 2 high. This bifurcation works very neatly to produce well-contrasted outcomes within the logic of the IL 2 × 2 framework.

However, *if* our scenarios are to provide targeted, broad-brush prediction, then a middle option may also be needed. We proposed the addition of a fifth scenario, in a model we call 'IL 5'. Here, we wish to convey a subtle distinction that positions the medium level of a critical uncertainty as something that is *probabilistically in the middle*, rather than as something that is *more probable*. Importantly, if this fifth scenario is to be a useful addition, it should not overshadow the rest and give us false confidence in a likely path.

Another key finding is that ostensibly practical decisions about scoping and framing of scenarios are shaped—in important ways—by the assumptions and values we hold. Which issues are in? Which are out? How are issues framed? Whose voices are heard, and whose are excluded? What types of data do we admit? How do we handle complexity? What expertise do we need? We argued that decisions about scenario process are not the objective, technical considerations they are often perceived to be. At the same time, in our experience, very few people within an organisation have the time or interest to consider abstract questions of epistemology and methodology—beyond the most basic considerations. A skilled, reflective facilitator may help. However, we suggest that putting some time and effort into prudent assessments and conscious choices when identifying our *objectives* (Level 1); *approaches* (Level 2); and *techniques* (Level 3) will yield a solid return on investment. If we do this, it is more likely that our scenarios will align with the sponsoring organisation's goals, and our outputs are more likely to be plausible, rigorous, and robust.

We observed that there are critical dynamics at work in the driving forces of geopolitical scenarios which are not always obvious. We suggested that unpacking these dynamics provides a more nuanced understanding of the causal impacts these driving forces are likely to have within a scenario. This can be useful in two ways. First, it may help clarify (and, if necessary, challenge) the assumptions that scenario participants may hold about the impact of a particular driving force. Second, it can help ensure that the decisions we make about the trajectory of a scenario narrative are internally consistent with our assumptions. Specifically, we noted that geopolitical driving forces tend to differ according to whether:

- they are *critical uncertainties* or *predetermined elements*;
- they *impel* events or trends forward or *constrain* them;
- they instigate *rivalry* or promote *cooperation*; and
- they are *exogenous* or *endogenous* to the field in question.

Further, we observed that these dynamics seldom occur as a simple binary choice: as totally uncertain OR totally predetermined. Instead, each driving force tends to be located somewhere *along a spectrum* within each of these dynamics. We also suggested that each driving force has an individual 'dynamic signature'. This is simply a graphical representation of where a driving force sits on each the four spectra. We suggest this offers a more accurate and nuanced picture of how each driving force is likely to behave.

Scenarios are a means for an organisation to run its current strategies through a 'wind tunnel' of contrasting but plausible future conditions, to test how resilient (or fragile) those strategies may be. The point is not to be totally prepared for one scenario, but somewhat prepared for all. However, we cannot develop scenarios that are plausible, rigorous, and robust (over the medium to long term) in a single afternoon. We echo Godet and Roubelat's warning that a good scenario process is technically demanding, time consuming and should take around a year (Godet and Roubelat 1996). We have argued throughout this book that, regardless of approach, committing time and resources to scenario development on this scale represents a very small fraction of the national investments directed to Antarctic programs.

A final observation on scenario analysis. We noted that the literature on scenarios is rich, diverse, and full of contradiction. This contributes to Martelli's now-famous critique that scenario planning is in a state of 'methodological chaos' (Martelli 2001, para 19). This has led some writers to lament that scenario theory lacks a common core, a coherent foundation. Indeed, we suggested that scenario planning suffers from a form of 'town and gown syndrome', in which its practice struggles for scholarly recognition and its theory eludes practical relevance. In part, this is because the practice of scenario planning has often developed in advance of theory. However, we argued that straining for a general or unified theory of scenario planning offends the principles of 'parsimony' in the construction of theory and minimalism in ontological commitments. In any event, we suggest that such a theory is possibly unattainable and probably unnecessary. As Martelli notes, practitioners have tended to just get on with it (Martelli 2001, para 26).

10.3 Our Findings: Geopolitical Futures for Antarctica

Turning our gaze to Antarctica, we found that scenario planning, in various guises, has underpinned significant research and logistics activity conducted by national Antarctic programs. We also saw that Antarctic futures have been the subject of serious academic attention. However, we noted that modelling of Antarctica's *geopolitical* futures is not nearly as comprehensive or well-resourced as it is for its *environmental* futures. It also appears that experts in geography, geopolitics, polar governance, and law seldom engage with formal scenario analysis. In this book, we argued that that a well-designed and executed scenario process can add to these disciplines, both by firing the imagination *and* being firmly tethered to a contemporary evidence base, with rich data analysis and robust chains of causal reasoning.

We found that, for the past 60 years, the Antarctic Treaty has underpinned what is widely acknowledged as one of the world's more successful regimes for regional governance. The Treaty provides the basis for peaceful use of the continent and its surrounding marine areas, fosters cooperation in scientific endeavours, and has spawned additional laws and institutions to manage marine resources and to protect the Antarctic environment. In doing so, it has addressed a range of internal and external pressures.

We noted that the 1980s and 1990s saw a profound shift in the internal direction of the ATS, with the Minerals Convention developed, then abandoned, then replaced with the Madrid Protocol. The ATS also faced external pressure from the relentless debates at the United Nations General Assembly over the 'Question of Antarctica'. The Antarctic Treaty Consultative Parties (ATCPs) successfully negotiated these stressors. In doing so, they managed to revitalise the regime's governance and its focus, with Antarctica designated as a 'natural reserve devoted to peace and science'.

Moving forward to the 2020s, we found the Antarctic geopolitical landscape is subject to some new driving forces—as well as many familiar ones. We presented a snapshot of these, grouped across nine themes: the standard PESTEL categories (Political, Economic, Social, Technological, Environmental and Legal) plus Sloan's three disciplines of geopolitics: Geographic, Strategic and Historical. We used these driving forces to define the premisses, and to shape the trajectories of our scenario narratives.

We noted that Article IV of the Antarctic Treaty has proven effective in balancing the rights and interests of claimants and non-claimant states. We suggested this 'bifocal' approach to the sovereignty issue remains pivotal to ongoing international cooperation on Antarctic matters. We also saw that Antarctica is regarded by some as a potential resource 'El Dorado', and that some believe it is only a matter of time before the current preservationist ethic in Antarctica is overwhelmed. We observed that the melting of sea ice due to climate change may open more areas to resource harvesting, such as fisheries, and human access, such as tourism.

We settled on the issue of *militarisation* to illustrate how scenarios of Antarctic geopolitics might be developed. Antarctica is commonly perceived as an area which has always been immune from military competition. However, as we noted, prior to the Antarctic Treaty, there had been a small but significant history of military activity and skirmishes. The bleak prospect of Antarctica becoming yet another theatre for potential hostilities between Cold War superpowers, or by claimant states, impelled Parties towards a peace agreement in the form of the Antarctic Treaty.

For 60 years, the Treaty has ensured that the ATA remains free from conflict, retains its non-militarised status, and has avoided the escalation of territorial disputes. The non-militarisation of Antarctica, which extends to nuclear weapons, also performs a strategic role *outside* Antarctica, in providing a stable, southern buttress for other regional agreements that ban nuclear weapons. We found that the Southern Ocean (including maritime approaches that extend into the ATA) remains strategically important for Antarctic gateway states. These states benefit greatly from the non-militarisation of Antarctica. It has enabled them to direct most of their defence capabilities toward other theatres of potential conflict. Another key driving force (this time in the form of a *constraint*) is Antarctica's geographic isolation and severe climate. This presents huge costs, logistical barriers, and risks for military activity on the continent—particularly in East Antarctica. We observed that some technologies being deployed within the ATA may be 'dual use', in the sense of being able to be seamlessly re-purposed between scientific and military capabilities. Such dual-use technologies include satellite receiving stations and infrared telescopes with capability to track satellites, missiles, and drones.

We suggested that the greatest risk here is the potential for dual-use equipment and personnel in the ATA to be re-purposed to support non-peaceful activities *outside* the ATAon land, in the oceans or in space. Further pressure may come from technological advances in ocean sensors, global positioning systems, telescopes, unmanned aircraft and ships and satellites. While the inspection systems created by the Antarctic Treaty and Madrid Protocol provide some level of compliance monitoring, the modest frequency of these inspections and the need for highly technical expertise to detect breaches mean that prospects of effectively policing this activity are slight.

We illustrated a process for constructing scenarios on Antarctic militarisation in the classical geopolitical tradition. We augmented the standard IL 2×2 process in several ways to address some of the difficulties encountered when applying this technique to Antarctic geopolitics. We identified around 50 driving forces relevant to Antarctic militarisation, grouped across nine themes. Of these, two driving forces stood out as the most uncertain and important. These *critical uncertainties* were:

Political	Stability of the international system, relevant to the ATS	**CU 1**
Strategic	Strategic advantage from Antarctic militarisation	**CU 2**

Using our 'IL 5' approach, we took these two critical uncertainties (at their high, medium and low values) to construct five scenarios on Antarctic militarisation.

In *Splendid Isolation*, ATCPs agree to a new transparency measure that acknowledges the changing nature of military technologies and reaffirms the importance of the Antarctic Treaty's prohibition on military activity. In *Quarantine,* a small number of powerful states challenge the rules and norms of the post-war international order—including the ATS. However, global tensions have not extended into the ATA. In *Stealth*, Antarctic Treaty processes have been largely unaffected great power rivalry, and the ATS continues to function—notwithstanding a lack of consensus over various measures for environmental protection. However, there is intense pressure to expand maritime surveillance systems, such as the 'Ocean of Things', into the ATA. In *Cold War II*, 'dual-use' equipment is deployed increasingly at Antarctic stations. Several major powers use a proxy state to pursue deep seabed mining within the ATA, in contravention of the mining ban. This activity is guarded by military equipment and personnel, culminating in a hostile incident.

Our fifth scenario, *Pressures Managed*, was the new medium option. This is designed to provide some evidence-based, targeted, broad-brush prediction, without suggesting this outcome is more probable. In this scenario, there are considerable tensions in the wider international system, but these are kept in check by a new regional balance of power. However, this is offset by steady growth in dual-use technologies; the reinterpretation of 'non-military' activity as meaning 'non-aggressive'; and use of the ATS consensus rules to block tighter controls on dual-use technologies.

From these five scenarios, we developed a set of policy-relevant recommendations. If the non-militarisation of Antarctica is to be preserved, two suggestions stand out. The first is that like-minded states should resist moves to reinterpret what constitutes 'military activity' in Antarctica to a lesser standard. The second is that the ATS should review its compliance and inspection regimes, to better account for contemporary military technologies. This exercise was an illustration, and we expect a richer set of findings would emerge from a 'real world' process, with a broader set of participants and multiple iterations.

Finally, we explored how scenarios on Antarctic militarisation might inform a classical geopolitical understanding of Antarctica and the ATS. We examined the intersections between spatial ordering, authority, and power in the subtle rivalry between states over Antarctica. Our scenarios illustrate how projection of power may influence the future of the ATS—both globally, from great power competition, and regionally, from simmering rivalries in the Indo-Pacific and South Atlantic. We suggested that we might view the balance of power not simply as an adversarial relationship, but also as an associational balance of power, with its potential for a just equilibrium between states. In turn, this enlivens the authority of enabling institutions (such as the ATS) and international law. We also saw that the spatial ordering of areas in, below and above Antarctica is subject to significant tension due to ambiguities in interactions between the United Nations Convention on the Law of the sea (UNCLOS), the Antarctic Treaty and treaties regulating outer space.

So, what should we make of all this? Clearly, the Antarctic Treaty founds a regime that promotes effective governance of a region described (evocatively, but accurately) in the Madrid Protocol as 'a natural reserve devoted to peace and science'. We hope to have shown that a classical geopolitical lens is important and relevant to the study of Antarctic futures. Our scenario approaches and techniques identified a number of key driving forces for change (impelling) and equilibrium (constraining). How well the ATS responds to these driving forces will turn on its resilience as a governance system. By this, we mean 'a capacity to prepare for, to respond to, or bounce back from problems or perturbations and disturbances, which cannot necessarily be predicted or foreseen in advance' (Chandler and Coaffee 2017, p. 4).

As we have seen, scenarios are useful in this zone beyond standard prediction—*provided* they are plausible, rigorous, and robust. It is our hope that like-minded Parties and researchers might collaborate in scenario work, to contribute to the resilience of the ATS in the challenging years ahead.

References

Chandler, David, and Jon Coaffee. 2017. Introduction. In *The Routledge Handbook of International Resilience*, ed. David Chandler and Jon Coaffee, 3–9. Abingdon, UK: Routledge.

Godet, Michel, and Fabrice Roubelat. 1996. Creating the Future: The Use and Misuse of Scenarios.*Long Range Planning* 29 (2): 164–171.

Martelli, Antonio. 2001. Scenario Building and Scenario Planning: State of the Art and Prospects of Evolution. *Futures Research Quarterly* 17 (2): 57–70.

Index

A

Aerial surveillance, 156, 174. *See also* unmanned aircraft
Ambiguity, 7, 44, 45, 53, 87, 115, 182, 192. *See also* TUNA conditions
Antarctica
 airspace, 181
 bioprospecting, 27
 Heroic Era exploration, 97
 militarisation, 1, 12, 71, 93, 121, 125–129, 136, 141, 143–145, 148, 149, 153, 161, 162, 164, 166, 176, 180, 182, 186, 191, 192
 mining ban, 94, 115, 135, 141, 166, 191
 non-militarisation, 12, 98, 105, 106, 109, 118–122, 135, 153, 157, 163, 165–167, 171–174, 177, 180, 182, 190, 191
 peaceful use, 135, 160
 Question of Antarctica (1982-2002), 25
 resources, 5, 99, 114
 science as a strategic activity, 5
Antarctic futures, geopolitical security framing, 6
Antarctic gateway states, 117
Antarctic Peninsula, 18, 20, 21, 97, 106–109, 113
Antarctic Problem, territorial claims, 20, 22
 standstill agreement (Escudero approach), 22
Antarctic Treaty (1959), seminal arms control treaty
 Article IV 'peaceful use', 23, 93, 113–115, 134, 172, 190
 Article V arms control, 111, 112, 120, 130
 Article VII pioneering inspection regime, 120, 130
 negotiations for, 23
 provisions of, 120, 165, 171, 172
Antarctic Treaty Area (ATA), 2, 3, 6, 12, 86, 93, 95, 98, 105, 106, 112, 113, 115, 117–119, 122, 125, 127–130, 132–134, 136, 145, 148, 154–158, 160, 162–167, 171, 172, 174, 176, 177, 180, 181, 183, 186, 190, 191
Antarctic Treaty Consultative Meetings (ATCMs), 3, 24, 25, 27, 28, 96, 97, 113, 154, 155, 157, 163–165, 173
Antarctic Treaty Consultative Parties (ATCPs), 3, 4, 25–27, 91, 93, 95, 96, 115, 155, 163, 166, 190, 191
Antarctic Treaty System (ATS)
 consensus mechanisms, 6
 dispute resolution mechanisms, 165
 exceptionalism of, 177
 international legal analysis, 7
 legitimacy and effectiveness, 91
 resilience, 5, 93, 192
Anticipatory systems theory, 39
Argentina, Argentine Antarctica, 18
Associational balance of power, 178–180, 182, 192
Astronomical and space research, 120, 122
Australia, *Antarctic 20-year Strategy and Action Plan* (2016)
 national Antarctic program, 1, 85, 111, 129, 137, 163, 173, 186, 189

J. McGee et al., *The Future of Antarctica*, Springer Polar Sciences,
https://doi.org/10.1007/978-981-16-7095-4

research stations, 22, 171
territorial claims, 1, 17–19, 21, 113–115
Autonomous sensors, 95
Authority, and legitimacy
compliance, 178, 179

B

Balance of power
adversarial versions of, 179
associational, 178–180, 182, 192
Balancing coalitions, 162, 177
Bandhold, Hans, 40, 61
Basket of scenarios
plausible, 43
probably, 43
possible, 43
Bergin, Anthony, 95
Black box disempowerment, 62, 64
Brady, Anne-Marie, 4, 86, 95, 118, 120
British Antarctic Territory, 18
Bull, Hedley, 179

C

Child, Jack, 86, 96
Chile, territorial claim in Antarctica
Chilean Antarctic Territory, 18
China, fisheries
13th Chinese Government Five-year Plan for the Economic and Cultural Development of the People's Republic of China (2016), 121
research stations, 163
Choke points
Malacca Straits, 136
Suez Canal, 136
Classical geopolitics, definitions, 8, 84
Clausewitz, Carl von, 173
Climate, 4–6, 8, 27, 84, 95–97, 105, 117, 128, 132, 133, 135, 140, 141, 144, 157, 163, 164, 176, 186, 190
Climate change, 4, 6, 95, 96, 117, 132, 133, 140, 144, 157, 163, 164, 167, 190
Cold War, 2, 20–24, 97, 106, 109, 111, 128, 190
Cold War II (scenario), 148, 149, 153, 158, 166, 172, 175, 177, 182
Command, Control, Communications, and Reconnaissance (CCCR), 128, 131, 132, 154, 155, 157, 159, 160, 163, 165, 166, 172, 181
Committee for Environmental Protection (CEP), 173
Common heritage of mankind, 25, 26, 99, 113
Compliance, 2, 4, 24, 117, 119, 120, 122, 134, 141, 153, 163, 165, 167, 174, 175, 178, 179, 191, 192. *See also* authority
Compliance and inspection regimes, 153, 165, 167, 192
Comprehensive Nuclear Test Ban Treaty (CTBT) 1996, international monitoring system (IMS), 112
Convention for the Regulation of Antarctic Mineral Resource Activities (CRAMRA), 3, 26
Convention on the Conservation of Antarctic Marine Living Resources (CCAMLR), 3, 24, 92
Council of Managers of National Antarctic Programs (COMNAP), 173
COVID-19, 4, 6, 118, 154, 157, 162, 185
Critical geopolitics, 8, 9, 12, 84
Critical realist geopolitics, 10
Critical uncertainties, 10, 36, 40, 52, 57, 69, 71, 72, 74, 76–78, 83, 89, 93, 96, 99, 125–129, 138, 142–146, 148, 149, 161, 162, 166, 167, 176, 180, 187, 188, 191

D

Deception Island, 107–109
Deductive approaches, intuitive
archetypal (utopian/dystopian), 57
normative, 53, 56, 57, 60, 67, 84, 144
Deep seabed mining, 26, 159, 166, 191
Derbyshire, James, 42, 88
Determinism, 8, 40, 60
Dodds, Klaus, 5, 97
Drake Passage, 118, 136
Driving forces, critical uncertainties
importance of, 95
predetermined elements, 10, 71, 83, 89, 99, 126, 129, 149, 188
Driving forces, dynamics of
constraining, 90
cooperation, 91
endogenous, 90
exogenous, 90
impelling, 90
rivalry, 91
Driving forces for militarisation in Antarctica, identifying
ranking, 129

scoring, 139
Driving forces, themes for
economic, 93
environmental, 95
geographic, 97
historical, 99
legal, 96
political, 93
social, 94
strategic, 97
technological, 95
Dual-use personnel, 106
Dual-use technologies, satellite receiving stations
infrared telescopes, 95, 120, 122, 132, 139, 143, 144, 154–156, 159, 160, 163, 164, 181, 190
Dynamic signatures, 89, 92, 146, 147, 149, 188

E
El Dorado, 93, 190
Engagement approaches, 65, 68, 70
Environmental Protocol to the Antarctic Treaty (Madrid Protocol), Article 14 (inspection)
review conference of, 115
Expectation value, 76, 161

F
Falkland Islands / Islas Malvinas, war, 2
Fisheries, 4, 85, 92, 93, 96, 117, 118, 130, 167, 190
Forecasting, 32, 181
Frame, Bob, 4–6
France, territorial claims in Antarctica, 109
Future, 177
Future-proofing, 76
Futures, 1–9, 11, 12, 21, 27, 28, 31, 32, 35–43, 45, 46, 53, 57, 59–62, 64–68, 70–72, 75–79, 83, 85, 87, 89, 90, 97–99, 106, 115, 116, 125, 127, 135, 138, 141, 144, 145, 149, 153–155, 157, 158, 166, 172, 175, 177, 179–183, 186, 187, 189, 192

G
Geography, 6, 8, 9, 84–86, 97, 105, 189
Geopolitics, classical
critical, 8
critical realist, 10
definitions, 8, 84
Germany, Antarctic expeditions
Neu-Schwabenland, 20
Pinguin, 107
Global Positioning Systems (GPS), 120–122, 131, 154, 156, 157, 163, 191
Godet, Michel, 31, 60, 187, 189
Great powers, competition, 17, 23, 28, 125, 128, 130, 145, 154, 155, 158, 162, 164, 166, 177, 178, 180, 182, 192

H
Hart, H. L. A., 174
Hawke, Robert (Bob), Australian Prime Minister, 26
Hazard of extrapolation, 41, 42
Hazard of invention, 41, 42
Hemmings, Alan D., 4, 93, 95, 97
High seas, freedoms
non-aggressive naval activities, 159
History, 8, 11, 12, 17, 27, 45, 84–87, 98, 105–107, 115, 119, 128, 164, 179, 186, 190
Hope Bay incident (1952), 109
Hughes doctrine, 19
Hybrid approaches, 58, 63, 64

I
Icebreakers, 108, 132, 140, 163
IL 2x2, 12, 52, 63, 69, 72, 76, 78, 89, 93, 138, 143–145, 147–149, 153, 161, 166, 187, 188, 191
IL 5, 52, 76, 78, 125, 126, 161, 188, 191
Illegal, Unreported, and Unregulated (IUU) fishing, 4, 116, 117, 133, 140, 156
Illusion of certainty, 43, 78
Imagination, 6, 35–37, 40, 41, 58, 94, 105, 145, 154, 189
Imperial Conference (1926), 19
India, 19, 21, 25, 27, 98, 99, 114, 116, 120, 155, 158, 159, 173, 177
Indo-Pacific 'Quad', 155
Indo-Pacific region, 6, 159, 162, 177
Inductive approaches, premisses, 56, 58, 60, 61, 72, 144
Institutional dynamics, 59, 67, 90
Institutionalist literature, 7

Integrated approaches, 64, 65
Intergovernmental Panel on Climate Change (IPCC), 5
International Association of Antarctica Tour Operators (IAATO), 7
International Geophysical Year (1957-58), 21, 22, 95, 108
International governance, 2, 7, 11, 17, 24, 27
International law, critical international law scholarship
  rational choice approach to, 178
Intuitive Logics (IL) school, 52
Intuitive thinking, 58, 59

J
Japan, Antarctic expeditions, 20
John Biscoe, 108
Joyner, Christopher, 7, 171, 172, 180
Just equilibrium of power, 179, 180

K
Kahane, Adam, 68
Kármán Line, 181
Kelly, Phil, 84, 86, 176, 177
Kerguelen Plateau, 163

L
La prospective, 60, 67, 68
Law, 2, 6, 7, 12, 24, 26, 60, 96, 111, 115, 126, 135, 141, 157–160, 162, 165, 167, 170, 175, 176, 178–182, 186, 189, 192
Lebensraum, 2, 86
Legal scholarship, traditional or doctrinal, 178
Legitimacy of rules, 175
Liggett, Daniela, 5
Lindgren, Mats, 40, 61
Little, Richard, 179
Logistics, 5, 22, 95, 97, 119, 157, 159, 163, 171, 172, 174, 189
Lukes, Steven, 170–172, 176

M
Mackinder, H. J., 85, 176
Madrid Protocol.*See* Environmental Protocol to the Antarctic Treaty
Malaysia, and the 'Question of Antarctica', 25, 99
Maps, 8, 10, 85, 109, 180
Marie Byrd Land, 108, 114
Marine living resources, 3, 98, 114, 116, 130, 135, 173
Marine monitoring systems, 121
Martelli, Antonio, 'methodological chaos', 46, 186, 189
Measures of a military nature, 98, 106, 111, 117, 119, 134, 141
Militarisation of Antarctica, driving forces of
  history of, 106
  military activity, 109, 110
  traditional military/strategic threats, 112
Military personnel and equipment, permitted roles of in Antarctica, 106, 112, 119, 122, 128
Minerals, Rare Earth Elements, 131, 139
Minerals, rare earth minerals, 162
Mining
  gas, 93, 163
  minerals, 93, 163
  oil, 93, 115, 163
Missile targeting, 95
Monroe Doctrine (1823), 85

N
Nation-state, 186
Newfoundland, 19
New Zealand, 156, 159, 171
Non-peaceful purpose, 119
Non-state governance, 7
Norway, 18, 20, 120, 127, 135, 173
Novelty, 44, 45, 87. *See also* TUNA conditions
Nuclear testing, 2, 27

O
Ocean observation systems, 122, 130, 132, 140
Ocean of Things, 157, 166, 191
Operation Highjump (1946-47), 108
Operation Tabarin (1943), 18, 108
Organisational learning, 38
Outer Space, 160, 165, 181, 182, 192
Outer Space Treaty 1967, 160, 165
Oxford Scenario Planning Approach (OSPA), stages of
  general approaches, 53, 54
  specific techniques, 53, 54
  strategic objectives, 53–55, 69, 70

P
Parsimony, theoretical, 46
Peaceful purpose, 119
Plausibility, and the 'aha' moment
attributes of, 40
distinct from probability, 40
evidence base for, 41, 42. *See also* zone of plausibility
Policy makers, 4, 60, 128, 180
Politics, economics, social, technological, environmental, legal (PESTEL) categories, 71, 83, 86, 190
Power, and international law
military power, 171
scientific and logistical power, 172
three faces of power, 171
Predetermined elements, 10, 71, 83, 89, 93, 99, 126, 129, 138, 142–144, 149, 188
Prediction, evidence-based
broad-brush, 42
by extrapolation of existing trends, 32
limits of, 31–34, 42, 45, 70, 187
targeted, 42–44, 51, 153, 162, 166, 187. *See also* zone beyond the limits of prediction
Preservationist ethic, 94, 190
Press, Anthony, 5, 117, 122
Pressures Managed scenario, 153, 162, 167, 181, 191

Q
Qualitative approaches, 58, 61–65
Quantitative approaches, 36, 54, 61–63
Quarantine scenario, 148, 153, 155, 166, 174, 177, 181, 191

R
Rainbow Warrior, 27
Ramírez, Rafael, 11, 40, 58, 59, 76
Regime theory, 7
Regression analysis, 62
Remote sensing, 95
Re-perceiving, 37
Resource exploitation, 94
Risk, 2, 31, 33, 40, 41, 45, 53, 55, 58, 64, 68, 71, 84, 88, 90, 96, 97, 105, 114, 122, 136, 165, 183, 187, 190, 191
Rivalry between states, 1, 9, 12, 17, 84, 91, 92, 169, 170, 176, 180, 182, 183, 192
Rocard, Michel, French Prime Minister, 27
Ross Dependency, 18
Ross Ice Shelf, 19, 21, 108
Roubelat, Fabrice, 60, 187
Royal Dutch Shell, 69
Rules-based order, 155, 158
Russia, Antarctic exploration, 17, 99
Russian Federation, 114, 171. *See also* Soviet Union

S
Satellite ground receiving stations, 120
Scenario, 'core team'
consultant, 56
matrix, 72
narratives, 75
planning, 1, 5, 6, 11, 31–33, 37, 39, 40, 42, 43, 45, 46, 52, 54, 63, 64, 76, 78, 85, 87, 106, 186, 187, 189
standpoint, 69
time horizon, 70
toolbox, 56
Scenario models, predictive models
quantitative models, 32, 61, 63
Scenario practice, aspects of (objectives, approaches, techniques)
framing before reframing, 54
fundamental assumptions of, 53
methodology, 56
preparatory steps, 54
Scenarios
complexity, 12, 31–34, 38, 39, 42, 44, 45, 51, 62, 70, 87, 99, 127, 187
imagination, 6, 36, 37, 58
normative, 6, 37, 60
passage of time, 31–34, 42, 45, 62, 70, 127, 187
uncertainty, 12, 31–34, 36, 37, 42, 44, 45, 62, 70, 72, 76, 87, 126, 127, 138, 177, 187
Scenarios for Antarctic militarization
Cold War II, 158
pressures managed, 162
quarantine, 155
splendid isolation, 154
stealth, 157
Scenarios, properties of
bounding the range of future possibilities, 35
capturing complex dynamics, 36
causal pathways to future states, 35
plausible stories about alternative future states, 35

reperceiving and challenging current assumptions, 37
structured thinking about alternative futures, 36
Scenarios, seven broad approaches to
deductive, 60
engagement, 65
hybrid, 63
inductive, 60
integrated, 64
quantitative, 61
scoping, 65
Scenarios, unified theory for
fundamental assumptions of, 53
Schmitt, Carl, 176
Schwartz, Peter, 38, 40, 71, 76
Science, collaboration
infrastructure, 163, 181
logistics, 163, 172, 174
Scientific Committee on Antarctic Research (SCAR), 5, 26, 94
Scoping approaches, 65
Second World War, 21, 22, 95, 106, 107. *See also* World War II
Sense of obligation, 174, 175
Sloan, Geoffrey, 8, 9, 83, 84, 86, 87, 177, 190
South Africa, 19, 26, 114, 116, 117, 127, 136, 156, 171, 173
South Atlantic, 98, 112, 113, 117, 180, 182, 192
Southern Ocean, shipping routes, 118, 122, 136, 156
Sovereignty, national
territorial, 4, 18, 23, 84, 134, 137, 169
Soviet Union, 20, 21–23, 26, 93, 97, 107, 109, 128, 131, 162. *See also* Russian Federation
Space (geographic), 175, 176. *See also* Outer Space
Spatial ordering of Antarctica, 181
Spatial rivalry, 10, 12, 70, 84, 91
Splendid isolation scenario, 148, 149, 154, 166, 174, 191
Spykman, Nicholas J., 85
State-centric, 10, 84, 127, 186
Stealth scenario, 148, 157, 166, 172, 175, 182
Strategic studies, 8, 84
Strategy, 1, 21, 31, 34, 38–40, 43, 46, 75, 76, 79, 86, 87, 115, 118, 126, 127, 164, 182, 186, 187
Surveillance drones, 118, 119

T
Third-party state, as proxy, 115, 116, 130
Tin, Tina, 5
Town and gown syndrome, 46, 189
Transparency measure, 155, 166, 191
Turbulence, 38, 44, 45, 87. *See also* TUNA conditions
Turbulence, Uncertainty, Novelty and Ambiguity (TUNA conditions), 44, 45

U
Uncertainty, 12, 31, 32–34, 36–39, 42, 44, 45, 51, 62, 70, 72, 76, 87, 92, 99, 125–127, 138, 142–144, 146, 148, 149, 177, 180, 187, 188. *See also* TUNA conditions
Uncertainty, definition, 33
United Kingdom, 'Heroic Era' exploration
territorial claims, 99
United Nations Charter (1945), 182
United Nations Convention on the Law of the Sea (UNCLOS), 96, 157, 181, 182, 192
United Nations General Assembly (UNGA), 21, 25, 26, 99, 113, 190
United Nations Moon Treaty (1979), 26. *See also* outer space
United States, Antarctic exploration
research stations, 1, 17–19, 22, 113–115
territorial claims, 1, 17, 18, 19, 21, 115
Unknown variables, causal structure of, 33
Unmanned aircraft and ships, 121, 122, 156, 191

V
Van der Heijden, Kees, 42, 58
Van Efferink, Leonhardt, 8, 10, 69, 86, 91

W
Wack, Pierre, the 'aha' moment, 39
Wilkinson, Angela, 40, 43, 44, 53–56, 58, 59, 76
Wind tunnel testing, 38, 75, 189
Woolcott, Richard, 26
World War II, 155. *See also* Second World War
Wright, George, 72, 75, 88

Y
Young, Oran R., 7

Z

Zone beyond the limits of prediction, 34
Zone of plausibility, 31, 42, 43, 46, 62, 75, 187
Zone of standard prediction, 43

www.ingramcontent.com/pod-product-compliance
Ingram Content Group UK Ltd.
Pitfield, Milton Keynes, MK11 3LW, UK
UKHW022211051026
15624UKWH00001B/107
9789811670947